Subband Compression of Images:
Principles and Examples

Series Editor: J. Biemond, Delft University of Technology, The Netherlands

ADVANCES IN IMAGE COMMUNICATION 6

Subband Compression of Images:
Principles and Examples

T.A. Ramstad
Norges Tekniske Høgskole
(Norwegian Institute of Technology)
Trondheim, Norway

S.O. Aase, J.H. Husøy
Høgskolen i Stavanger
(Rogaland University Center)
Stavanger, Norway

1995

ELSEVIER
Amsterdam – Lausanne – New York – Oxford – Shannon – Tokyo

ELSEVIER SCIENCE B.V.
Sara Burgerhartstraat 25
P.O. Box 211, 1000 AE Amsterdam, The Netherlands

Library of Congress Cataloging-in-Publication Data

Ramstad, T. A.
   Subband compression of images : principels and examples / T.A.
Ramstad, S.O. Aase, J.H. Husøy.
      p.   cm. -- (Advances in image communication ; 6)
   Includes bibliographical references adn index.
   ISBN 0-444-89431-4 (alk. paper)
   1. Image processing--Digital techniques. 2. Image compression-
-Data processing. 3. Coding theory.   I. Aase, S.O. II. Husøy, J.
H. III. Title. IV. Series.
TA1637.R36   1995
621.36'7--dc20   95-17786
                                                         CIP

ISBN: 0444 89431 4

This book is printed on acid-free paper.

Transferred to digital printing 2006

Printed and bound by Antony Rowe Ltd, Eastbourne

INTRODUCTION TO THE SERIES
"Advances in Image Communication"

Image Communication is a rapidly evolving multidisciplinary field focussing on the evaluation and development of efficient means for acquisition, storage, transmission, representation and understanding of visual information. Until a few years ago, image communication research was still confined to universities and research laboratories of telecommunication or broadcasting companies. Nowadays, however, this field is also witnessing the strong interest of a large number of industrial companies due to the advent of narrowband and broadband ISDN, digital satellite channels, digital over-the-air transmission and digital storage media. Moreover, personal computers and workstations have become important platforms for multimedia interactive applications that advantageously use a close integration of digital video compression techniques (MPEG), Very Large Scale Integration (VLSI) technology, highly sophisticated network facilities and digital storage media. At the same time, the scope of research of the academic environment on Image Communication has further increased to include model- and knowledge-based image understanding techniques, artificial intelligence, motion analysis, and advanced image and video processing techniques and lead to a diverse area of applications such as: access to image data bases, interactive visual communication, TV and HDTV broadcasting and recording, 3D-TV, graphic arts and communication, image manipulation, etc. The variety of topics on Image communication is so large that no-one can be a specialist in all the topics, and the whole area is beyond the scope of a single volume, while the requirement of up-to-date information is ever increasing.

In 1988, the European Association for Signal Processing EURASIP together with Joel Claypool & Ir. Hans van der Nat, at that time Publishing Editors at Elsevier Science Publishers, conceived several projects to meet this need for information. First of all a new EURASIP journal, "Signal Processing: Image Communication", was launched in June 1989 under the inspired Editorship of Dr. Leonardo Chiariglione. So far, the journal has been a major success not in the least due to the many special issues devoted to timely aspects in Image Communication, such as low/medium/high bit rate video coding, all digital HDTV, 3D-TV, etc. It was further decided to publish a book series in the field, an idea enthusiastically supported by Dr. Chiariglione. Mr. van der Nat approached the undersigned to edit this series.

It was agreed that the book series should be aimed to serve as a comprehensive reference work for those already active in the area of Image Communication. Each volume author or editor was asked to write or compile a state-of-the-art book in his area of expertise, and containing information until now scattered in many journals and proceedings. The book series therefore should help Image Communication specialists to get a better understanding of the important issues in neighbouring areas by reading particular volumes. At the same time, it should give newcomers to the field a foothold for doing research in the Image Communication area. In order to produce a quality book series, it was necessary to ask authorities well known in their respective fields to serve as volume editors, who would in turn attract outstanding contributors. It was a great pleasure to me that ultimately we were able to attract such an excellent team of editors and authors.

The Series Editor wishes to thank all of the volume editors and authors for the time and effort they put into the book series. He is also grateful to Ir. Hans van der Nat and Drs. Mark Eligh of Elsevier Science Publishers for their continuing effort to bring the book series from the initial planning stage to final publication.

Jan Biemond
Delft University of Technology
Delft, The Netherlands
1993

Future titles planned for the series "Advances in Image Communication":

– HDTV Signal Processing	R. Schäfer, G. Schamel
– Magnetic Recording	M. Breeuwer, P.H.N. de With
– Image Deblurring; Motion Compensated Filtering	A.K. Katsaggelos, N. Galatsanos
– Colour Image Processing	P.E. Trahanias, A.N. Ventsanopoulos

Preface

Subband coding for compression of images is a rather modern technique. While it has been popular in speech and audio coding for more than a decade, its use in image coding is more recent. Consequently, much of the early work on subband coding of images has directly exploited the experience gained in audio applications.

Transform coding and subband coding were originally developed separately, but the newer of the two, subband coding, has benefited considerably from transform coding theory and practice. The acknowledgment that both methods are closely related has been long prevailing, but the two coding communities have maintained their own terminology and views. An example of this is the *lapped orthogonal transform* (LOT) which originated from the transform community. The LOT was originally introduced as an extended type of transform operating on overlapping blocks. From a subband point of view, the LOT is considered as a bona fide *filter bank* with a firm theoretical foundation – a foundation which has not always been exploited in the subband community.

In this book, although the title indicates that we shall exclusively deal with subband coding issues, we make no principal distinction between subband coding and transform coding, but rather try to unify the two approaches. Similarities and differences are pointed out when appropriate. We do so by applying tools such as series expansion, elements from linear algebra, as well as multirate digital signal processing. The generality of this theory is such that both transform and filter bank decompositions, as well as differential coding, can be fit into a general framework. In addition to providing insight, the use of these tools enables us to formulate optimization problems whose solutions lead to superior image coders.

In large parts of the theory presented in this book, no explicit distinction between audio (one-dimensional), still-image (two-dimensional), and video (three-dimensional) compression will be made. All the theory that is most conveniently expressed in terms of one-dimensional signals can easily be extended to separable two- or three-dimensional signals and systems. Non-separable systems will not be treated explicitly in this book, but mention of such systems will be made when appropriate.

The *wavelet transform* has recently been introduced as a new signal decomposition technique for use in image coding. We hold the position that the wavelet transforms are nothing more than particular instantiations of subband filter banks. Consequently, their use does not introduce anything significantly new in an image coding context except for providing a different way of understanding some concepts. For this reason the wavelet transform has not been given any attention in this book.

It is our hope that this volume will contribute to the understanding of frequency domain coding techniques. Many images from coding experiments are presented for the purpose of making it possible for the reader to make judgements on the properties of different coders. We do not pretend to give a complete overview of all the work that has been done on frequency domain coders and in particular subband coders. Given the enormous activity in this field, that would be an overwhelming task. Also, our aim has been to present our own personal views on image subband compression and related issues. We have included what we think is important. This selection is of course somewhat colored by our own inclinations. This does, of course, not imply that the issues that we do not cover are unimportant or not worthy of mention.

Outline of the book: The organization of the book is as follows: In Chapter 1 we introduce the problem of image compression in general terms. Sampling of images as well as several other fundamental concepts such as entropy and the rate distortion function are very briefly reviewed. The idea of viewing coding techniques as series expansions is also introduced. The following chapter presents signal decomposition and the conditions for perfect reconstruction from minimum representations. Chapter 3 deals with filter bank structures, mainly those having the perfect reconstruction property. Subsequently, in Chapter 4 quantization techniques and the efficient exploitation of the bit resources are treated. In Chapter 5 we develop gain formulas – i.e. quantitative measures of the performance of filter banks in a subband coding context, and employ these in the search for optimal filter banks. While Chapter 4 deals with the utilization of bit resources from a theoretical perspective, Chapter 6 continues with this issue, but now from a more practical point of view. A number of examples of coded images using different subband coders are presented in Chapter 7. The aim of this chapter is to make it possible for the reader to make his own judgments as to the quality and properties of various types of image subband coders. From the examples presented in this chapter it will be evident that subband coders in general give rise to some characteristic types of image degradations. Therefore, in Chapter 8, we present several techniques for minimizing these artifacts. Finally, in Chapter 9, the theory and practice of subband coding of video, at several target bit rates, is presented.

How to read this book: When writing this book we have tried to organize the material in a logical progression. As such, the book can in principle be read sequentially. Realizing that the readers of this book will have a largely nonuniform background, we have also tried to make the various parts of the book as self-contained as possible. For example, Chapter 9 can be read almost independently from the other parts of the book. Also, Chapter 3 is more or less a self-contained treatment of filter bank theory using z-domain analysis techniques. The sequence of Chapters 2, 5, 8 are most suitably read in that order. The same statement can be made for the sequence of Chapters 4 and 6. We also envisage that some readers may wish to browse through the book to read only selected parts. It is our hope that this book can be benefited from if read in any of these ways.

Acknowledgments: The work on this book has taken much longer time than any of us expected. This has put strain on our working environments as well as on our families. We are grateful that our colleagues have tolerated reduced efforts in other endeavors during the most active writing periods. And we are forever indebted to our wives, Marie, Inger, and Johanne, who suffered more than us during the long creation process. Whereas we – from time to time – experienced the satisfaction of creating something, their dubious plight was to endure a mixture of our infantile exaltations and recurrent grumblings.

Although most of the strain has rested on our weak shoulders, we want to thank a selected group of people for contributions. John Markus Lervik has contributed significantly to Chapter 4, made valuable critical comments to other parts of the book, and acted as an excellent proof reader of whatever material was given to him. Likewise, Geir Øien's vigilant eye caught mistakes that nobody else would have thought possible. Finally, we thank Ilangko Balasingham and Morten Høye for providing some of the filter optimization results.

Trondheim/Stavanger, January 1995

Communis error facit jus

Contents

xvi

Chapter 1

Introduction

The necessity of image compression is becoming ever more evident. There seems to be a never ending increase in the amount of image and video material that must be stored and/or transmitted. Due to the rapid development and proliferation of digital storage media and the introduction of digital telecommunications networks such as the Integrated Services Digital Networks (ISDN), it is natural to store and transmit both still images and video using digital techniques.

Without compression the digital representation of an analog signal requires more bandwidth for digital transmission or storage than would the original analog representation. The combination of digital compression and digital transmission/storage is therefore a necessity in numerous applications.

When attempting to compress digital images several questions naturally arise:

1. What is it that makes image compression possible?

2. How is image compression performed?

3. By how few bits can we represent an image digitally while maintaining acceptable visual quality?

This book focuses on the second question and considers the class of source coders referred to as frequency domain coders. Our aim is to contribute to the basic understanding of the mechanisms that make image compression

1

possible, present the theory for subband coding and the relationships between this coding technique and other coding methods, as well as presenting results that indicate possible practical limits of important state-of-the-art coding techniques.

Transform coding has been, and still is, the dominant technique in existing image and video compression standards and implementations. The success of transform coding schemes can be attributed to the simplicity of their implementation, the sound theoretical foundation on which they are based, as well as their, for many practical purposes, good performance. Subband coding may be viewed as a generalization of transform coding and offers more flexibility through the availability of more free design parameters. Thus, we will stress the similarities and relationships between transform and subband coding techniques: Both operate in the frequency domain and obtain data compression through the same basic technique, namely through the distribution of bits to the different frequency bands/transform coefficients based on their relative importance. Traditional subband coding is computationally more complex than transform coders, but as will become evident later on, filter banks – the main building block of a subband coder – of lower computational complexity than transforms do exist. Also, filter banks with a higher count of arithmetic operations than a transform are in many instances suitable for VLSI implementation. The advantages of subband coders over transform coders are twofold:

- The theoretically obtainable compression ratio (as measured by the quantity *coding gain* to be defined later) is somewhat higher.

- Also, blocking artifacts, a typical distortion introduced by transform coders, can be completely avoided in subband coders.

On the other hand, ringing artifacts in the vicinity of sharp edges in the image, is an artifact that is characteristic of subband coded images at low bit rates. In this book we show that by combining the advantages of the two coder families we can design image subband coders with negligible blocking *and* ringing artifacts.

In video coding the situation is more complex. In this case the subband decomposition is typically combined with *differential* coding in a hybrid scheme involving motion compensation along the time-axis. When used in this context the salient features of a subband approach have a less well-founded theoretical basis. Nevertheless, several successful subband schemes for video coding targeted at low, medium, and high bit rates are known.

In the remaining part of this introductory chapter we present some useful background material. Following a brief introduction to image sampling/digitization, we discuss at some length the mechanisms that make image compression possible. The chapter is closed by an introduction to the *series expansion* approach, a generic description of various coding schemes.

1.1 Image Digitization

In this book we will invariably assume that the image signals to be compressed are band-limited analog signals sampled at or above the Nyquist rate. In such cases the sampled signal is an exact signal representation according to the sampling theorem [1]. The image samples are called picture elements, pels, or pixels.

If the original image, in its analog representation, is not band limited, a lowpass anti-aliasing filter is necessary to ensure that the signal is properly band limited prior to the sampling process. Following the sampling of the analog image signal, the amplitudes of the space-discrete signal are quantized and represented by a finite number of bits. Typically, the samples are organized in a rectangular array with each element, in the case of monochrome images, representing the luminance at a given point in the image. This rectangular array can be thought of as a matrix. Each luminance value is commonly represented by an 8 bit number. In the case of color images three 8 bit numbers are often used to represent the luminance component and two chrominance components. An 8 bit representation for monochrome images suffices for most applications and viewing conditions. However, for certain classes of images, such as X-ray images, more bits are required due to the large dynamic range of standard X-ray images on negative film.

In most circumstances digital images are formed by spatial sampling of the analog image on a rectangular grid as illustrated in Figure 1.1 a). Other spatial sampling patterns are also possible and is used for example in digital television. An example of such a sampling grid is the *quincunx* pattern which is illustrated in Figure 1.1 b). In this book we will almost invariably be concerned with digital images sampled on a rectangular grid. The reader interested in more material on alternative sampling patterns and processing of the corresponding digital images is referred to [2, 3, 4].

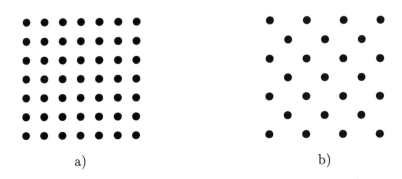

Figure 1.1: *a) Rectangular and b) quincunx sampling of images.*

1.2   Image Compressibility

As stated before, there are three fundamental questions associated with the task of image compression. After addressing the question of *"what makes image compression possible?"*, we present a discussion on some of the other fundamental issues.

1.2.1   What makes Image Compression Possible?

There are two main reasons why images can be compressed:

- All meaningful images exhibit some form of internal structure. When considering digital images as realizations of two-dimensional stochastic processes, this structure manifests itself through statistical dependencies between pixels. We refer to this property as *redundancy*.

- The human visual system is not perfect. For this reason certain image degradations are not perceivable. We refer to the allowable degradation due to shortcomings in the human visual system as *irrelevancy* or *visual redundancy*.

The two key parameters for evaluating a coding scheme are the *bit rate* and the *distortion* of the compressed representation. By bit rate we mean the average number of bits per pixel required for the representation of an image. The distortion is a numeric measure of the difference between two images. In a coding context it is most often used as a measure of the difference between the image to be coded and the image reconstructed from its compressed representation. Obviously, the lower the bit rate, the lower is the cost associated with the storage and transmission of the compressed image. On the other hand, as the bit rate goes down the distortion tends to increase. Thus, when applying image compression for some application we are faced with the conflicting requirements of simultaneously low bit rate and low distortion.

1.2.2 Distortion Measures

Distortion is a measure of the quality of a compressed representation of some signal. Viewing the signal to be compressed as a realization of a stochastic process it is common practice to use the mean square error

$$D = E[|X - \hat{X}|^2] \tag{1.1}$$

as a measure of the distortion between the original signal (represented by the stochastic variable X) and the signal reconstructed from the compressed representation ($\hat{X}$). In evaluating practical image coding algorithms, the *(Peak-to-Peak) Signal to Noise Ratio* (SNR) is often used. The SNR is defined for a signal with M samples by

$$SNR = 10 \log_{10} \frac{x_{max}^2}{\sum_{l=0}^{M-1} (x(l) - \hat{x}(l))^2}, \tag{1.2}$$

where $x(l)$ are the original signal samples and $\hat{x}(l)$ are the reconstructed samples. When coding monochrome images the maximum amplitude x_{max} is usually 255 since the allowable amplitude levels are 0 to 255 for 8 bit images.

The SNR is an example of a *Mean Square Error* (MSE) distortion measure. MSE distortion measures have enjoyed widespread popularity because of their conceptual simplicity and also because their use makes it reasonably easy to perform quantitative distortion calculations and optimizations.

Given the practical utility of MSE based distortion measures, it is still pertinent to ask if these are good measures of visual degradation. We indicate an answer to this question by means of an example. Keep in mind that it is highly desirable that two images that are visually similar should have little distortion between them. Suppose we have an image or a portion of an image containing stationary, white Gaussian noise. The visual appearance would hardly change if we replace this image part by another set of noise samples with the same statistics (such as mean value and variance) as illustrated in Figure 1.2.

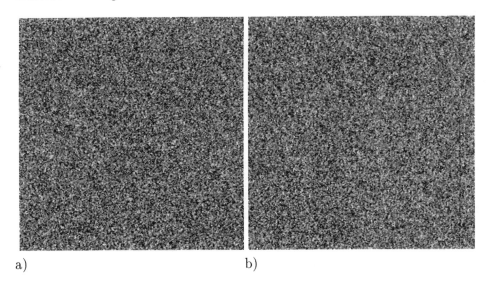

a)   b)

Figure 1.2: *Gaussian noise images. The SNR between the two image parts is 9.4 dB. Can you see the difference between them?*

In the above example (Figure 1.2) we notice that the SNR between the two noise-like images is low even though the images are visually very similar. Also, if the distortion between one image and the same image with a small constant shift in gray-level values were computed, we would get a small SNR even though the images would be virtually indistinguishable under normal viewing conditions. Other examples of large distortion, in a mean square sense, between visually indistinguishable images can be conceived.

It is also easy to construct examples in which we have identical images except in small localized areas where the distortion between them is unacceptably high for a human viewer. Nevertheless, we would get a high SNR since the error is localized. These examples illustrate some of the weaknesses of single number MSE based distortion measures. In general, it would be highly surprising if we could express such a complex property as image quality through a single number, irrespective of distortion measure.

Before closing this subsection we would like to point out that, in spite of all their weaknesses, the MSE based distortion measures have great practical utility for example when evaluating parameter selection for a given coding algorithm or when evaluating a selection of coding algorithms that are closely related. Also, the concept of distortion represented as a scalar function of the difference between the original and reconstructed signal is powerful in practice.

1.2.3 Visual Entropy

Bit rate – or *rate* for short – is related to the information content in the signal. The information content is measured by the *entropy*. For *discrete* amplitude signals, the bit rate required for its representation as well as its entropy is finite. In the following we shall review the concept of entropy for discrete amplitude sources and use it when presenting some philosophical arguments about the compressibility of digital images.

Assume that a discrete source generates symbols $x(l)$. These may be discrete amplitude pixel values of a digital image. N signal samples can be arranged in a vector $\mathbf{x}$. If a particular vector $\mathbf{x}$ has the associated probability $P_{\mathbf{x}}(\mathbf{x})$, then the per symbol entropy, in bits per symbol, for the elements in the vector is [5]

$$H_N(X) = \frac{1}{N} E[-\log_2 P_{\mathbf{x}}(\mathbf{x})] = -\frac{1}{N} \sum_{\forall \mathbf{x}} P_{\mathbf{x}}(\mathbf{x}) \log_2 P_{\mathbf{x}}(\mathbf{x}), \qquad (1.3)$$

where $E[\cdot]$ is the expectation operator. The average source entropy per symbol is then defined by the limit

$$H(X) = \lim_{N \to \infty} H_N(X). \qquad (1.4)$$

Finding this limit for a particular source is not easy in practice because of the difficulties in estimating the probabilities involved when N is a large number. In the following we use the entropy as a measure of the number of bits required for the representation of symbols from a given source.

Given a distortion measure, it is of utmost importance to find the lowest attainable bit rate for a given distortion. Every practical coding method could then be evaluated against such a limit. Such a function is the socalled *distortion-rate function* [6]. An example of a distortion-rate function is given in Figure 1.3.

Assume that we can define a subjectively correct distortion measure and that the obtainable bit rate can be estimated by the entropy. We can then

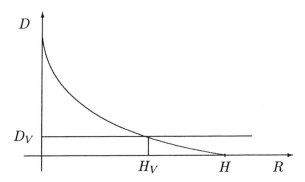

Figure 1.3: *Distortion-rate function of a discrete amplitude source. The threshold D_V is the lowest observable distortion, and H_V is the corresponding entropy, which we denote visual entropy.*

find a particular rate above which the distortion is not perceived by a human observer. We refer to this rate as the *visual entropy* of the source.

We now wish to put the concept of visual entropy into more precise terms. For this purpose, let us define the collection of all possible digital images of a given spatial extent and a given number of bits for the representation of each pixel as the *image set*. A large portion of the images in this set of all possible images would be noise-like images. From the previous discussion it is likely that they are visually indistinguishable if their stochastic characteristics are close. We would therefore expect that the subsets containing visually indistinguishable or very similar images would be rather large. Also, most real images would be highly unlikely to occur and would therefore not contribute much to the entropy of the source generating images in the *image set*.

Let the number of images in the *image set* be N_S. As an example, the set of images of extent 512×512 pixels with each pixel being represented by 8 bits has $2^{512 \times 512 \times 8}$ elements. This is, no doubt, a very large number. Denote by P_i the probability of each point in the image set. The entropy is given by

$$H_S = - \sum_{\forall i \in N_S} P_i \log_2 P_i. \tag{1.5}$$

With our assumed perceptually correct distortion measure it is obvious from our previous deliberations that images within reasonably large subsets of the image set are visually indistinguishable. Thus, if we can partition the image set into a collection of disjoint subsets each containing visually indistinguishable images, it is reasonable to use *one single element in each disjoint subset as the representative of the whole subset.* Such a partitioning of the image set along with the representative image is illustrated in Figure 1.4. The collection of these representative elements, or images, con-

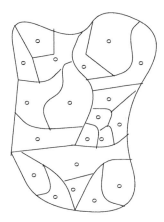

Figure 1.4: *Visualization of the image set. Each region contains visually different images, but all images within one region are indistinguishable and can thus be represented by a single image as illustrated by the circles.*

stitutes a subset of the image set. We call this subset $\mathcal{R}$. Assume that there are $N_{\mathcal{R}}$ images in this subset and that image i appears with a probability $\hat{P}_i$. Given this, the visual entropy can be defined by

$$H_V = - \sum_{\forall i \in N_{\mathcal{R}}} \hat{P}_i \log_2 \hat{P}_i. \tag{1.6}$$

The minimum attainable bit rate is lower bounded by this number for image coders without visual degradation.

Using the above model, lower bit rates are attainable by allowing visually *observable* distortion. This corresponds to enlarging the size of the subsets into which the image set is partitioned. An obvious consequence of this is that the number of subsets in the partition becomes smaller, the number of representative images – one for each partition – decreases, and the entropy of the collection of representative images decreases.

The defined visual entropy is a means of creating some general notion of limits to coding. To make it useful, we must define distortion measures that incorporate properties of the human visual system, perform the image set partitioning, and finally calculate probabilities for the respective regions. To achieve the first task we need more research on the human vision to obtain better visual perception models. Successful partitioning of the image space depends on the simplicity of the distortion measure. Even for simple distortion measures – given the dimension of each element of the image set – this is obviously an impossible task.

1.2.4   The Ultimate Image Coder

From a theoretical point of view, knowing a subjectively appropriate distortion measure, we can construct a coder with no visual degradation operating at a bit rate given by the visual entropy by using *vector quantization* [7]. Below we outline the principle for such an optimal coder.

All the representative elements (images, or vectors) of each of the disjoint partitions of the image set are stored in what is usually called a *codebook*. Both the encoder and decoder have copies of this codebook. When coding a particular image, it is compared to all the vectors in the codebook applying the visually correct distortion measure. The codebook entry with the closest resemblance to the sample image is the image corresponding to its compressed representation. The index (address) corresponding to this codebook entry is transmitted to the decoder as the image code. The decoder finds the correct representation as the address given by the transmitted index. Since the entries in the codebook are not all of equal probability, entropy coding should be applied to the addresses prior to transmission. In entropy coding probable indices are represented using few bits, whereas less probable indices must use more bits. The average bit rate is thus minimized.

Obviously, the method is unrealistic even if we knew the correct visual distortion measure and we were able to partition the space according to this measure. The complexity problem would be beyond any practical limit both with regard to the storage requirement for the codebook as well as in terms of computational complexity. We should therefore only view the indicated coding strategy as a limiting case. All practical coding schemes would be inferior.

1.3 Practical Approaches to the Coding Problem

In practical coding methods there are basically two ways of avoiding excessive storage and complexity demands of the above strategy.

- The first method is still based on pure vector quantization, but smaller vector sizes are used. This is often done by partitioning the image into a collection of smaller subimages. The strategy outlined above is then applied to each subimage. Doing this is obviously suboptimal since we can no longer exploit the redundancy offered by large structures in an image. But the larger the blocks, the better is the method. A further reduction of the processing burden is obtained by introducing some structure into the codebook that makes it possible to perform the comparison between the vector to be coded and the members of the codebook faster. The second problem, large memory for codebook storage can be alleviated by storing only parts of the codebook and use algorithms that can generate the other images from the stored ones. This is the technique used in *fractal* coders where the purest version avoids all codebook storage [8].

- The second strategy is is to apply some preprocessing on the image prior to quantization. The aim is to remove statistical dependencies among the image pixels. In the extreme case where all dependencies are removed, simple *scalar* quantizers can be efficiently used. To put this in the vector quantization terminology, the vector size is reduced to one pixel.

In the following the main emphasis will be on the second class of techniques although occasionally combinations with vector quantization will be employed. The basic idea in subband/transform coding is to split the image into subbands/transform coefficients with little correlation across the bands/coefficients. This is followed by some form of scalar quantization in which important subbands/transform coefficients are represented with many bits and the less important ones are represented with few or no bits at all. Both steps in this process are equally important. The role of the subband/transform decomposition is to create an image representation with as little crossinformation between the components as possible. The quantization and possibly subsequent entropy coding perform the actual compression of the image information.

In the next section we describe the commonly used image decomposition techniques. Series expansion will be used as a unifying concept both in that section and in the chapter to follow.

1.3.1 Coder Models

To understand similarities and differences between various coding methods, it is beneficial to introduce a general model that can appropriately describe most methods. It turns out that it is easier to construct a common model for the decoders than for the encoders. The model would typically consist of three parts as shown in Figure 1.5. The first block in Figure 1.5 receives the

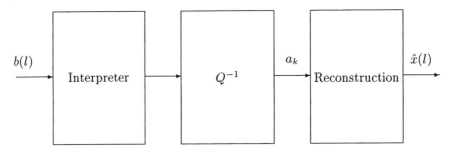

Figure 1.5: *Block diagram of generic decoder structure.*

bit-stream, as created for example by a Huffman coder, which it translates into indices specifying a representative value out of the inverse quantizer. These representative values are approximations to parameters that are necessary to produce a reconstructed version of the coded image. The nature of the parameters $\{a_k\}$ of Figure 1.5 depends on the exact coding method employed. For most coding methods the *reconstruction* box of Figure 1.5 can be expressed as a synthesis formula in the form of a *series expansion*

$$\hat{x}(l) = \sum_k a_k \phi_k(l), \tag{1.7}$$

where $\{a_k\}$ is the coefficient set (the parameters in the representation) and $\{\phi_k(l)\}$ are the basis functions. The series expansion concept serves as our unifying concept.

The most notable difference between compression schemes is the choice of basis functions, and the way in which the coefficients are represented.

For clarity of presentation we will restrict the discussion to one-dimensional signals. The results can easily be extended to two-dimensional separable systems. Generalizations to two-dimensional *nonseparable* systems are also possible [3, 4, 9].

In the remainder of this chapter we briefly review some basic coding techniques using the series expansion concept. In the next chapter we elaborate on the series expansion representation for subband and transform coders.

Differential Coding

Basic block diagrams of a one dimensional *closed-loop differential* encoder and decoder are shown in Figures 1.6 a) and 1.6 b), respectively. The de-

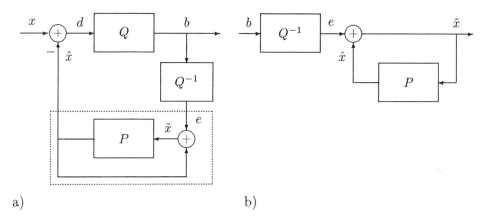

a)   b)

Figure 1.6: *a) DPCM encoder. b) DPCM decoder.*

coder consists of an inverse quantizer, Q^{-1}, and linear predictor, P. Such a predictor can be viewed as an FIR filter, thus we denote it by $P(z)$. The predictor makes a prediction of the next reconstructed sample, $\hat{x}(l)$, based on previously decoded samples, $\tilde{x}(l-1), \tilde{x}(l-2), \ldots$. The next reconstructed sample is constructed according to

$$\tilde{x}(l) = \hat{x}(l) + e(l). \tag{1.8}$$

Notice that the feedback loop in the decoder involving the predictor $P(z)$ is an IIR filter, let us call it $G(z)$, which is given by $G(z) = 1/(1 - P(z))$. The corresponding unit pulse response $g(l)$ is given by the inverse z-transform of $G(z)$. Notice also that the same filter, $G(z)$ is present in the encoder, see the dotted box in Figure 1.6 a). In the encoder the reconstruction filter $G(z)$ in the feedback loop estimates the next input sample based on quantized versions of the difference signals $d(l)$. The block denoted Q is the quantizer which outputs the quantizer index $b(l)$ which identifies the correct recon-struction value in the inverse quantizer Q^{-1}. The cascade of Q and Q^{-1} produces an approximation to the quantizer input. They are not exact in-verse operators as the symbols indicate. The signal $\tilde{x}(l)$ is the receiver signal approximation which will be equal in the encoder and the decoder provided that no transmission errors occur. The reason why differential coding, as explained above, is advantageous compared to straight forward scalar quan-tization of the original signal is that the dynamic range of $d(l)$ typically is much smaller than the dynamic range of $x(l)$. Thus, fewer bits are needed for its accurate representation.

The predictor and quantizer can be fixed or adaptive. The predictor adaptation can be based either on the incoming samples, in which case the predictor coefficient code has to be transmitted as side information along with the prediction error, or it can be computed from the quantized output samples, which makes the receiver able to derive the prediction coefficients from the incoming bit-stream. Also the quantizer can be made adaptive. More details on the many different differential coding structures can be found in [10].

An alternative description of the reconstruction process to that given above is as follows. Consider a fixed predictor. The output signal, $\tilde{x}(l)$, is given as the convolution between $e(l)$ and $g(l)$. Thus,

$$\tilde{x}(l) = \sum_{k=-\infty}^{\infty} e(k)g(l-k). \qquad (1.9)$$

The signal reconstruction can be viewed as a series expansion of the out-put signal consisting of basis functions generated by shifts of a single basis function (the unit sample response $g(l)$) and with the coded difference sig-nal $e(k)$ as coefficients. To more clearly see that the output signal can be viewed as a linear combination of (basis) vectors, we write out Equation 1.9 explicitly in vector notation. In other words, using Equation 1.9 to find

$[\ldots, \tilde{x}(0), \tilde{x}(1), \ldots]^T$, we get

$$
\begin{bmatrix} \vdots \\ \tilde{x}(0) \\ \tilde{x}(1) \\ \tilde{x}(2) \\ \vdots \end{bmatrix} = \ldots + e(0) \begin{bmatrix} \vdots \\ g(0) \\ g(1) \\ g(2) \\ \vdots \end{bmatrix} + e(1) \begin{bmatrix} \vdots \\ g(-1) \\ g(0) \\ g(1) \\ \vdots \end{bmatrix} + \ldots \qquad (1.10)
$$

We have illustrated the basis functions in Figure 1.7 for the case when $g(l)$ is causal.

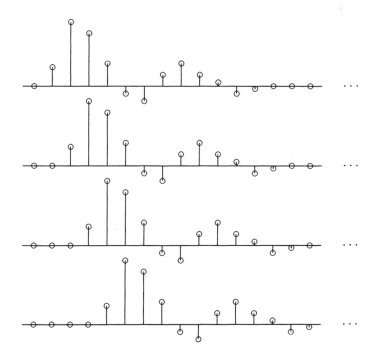

Figure 1.7: *Example of basis functions in the DPCM decoder.*

An *open-loop DPCM* (also called D*PCM) decoder has the same structure as the DPCM decoder, but the encoder filter $G(z) = 1/(1 - P(z))$ is
in this case put prior to the quantizer, as shown in Figure 1.8. The optimal
predictor is different in this case, as will be discussed in Section 5.2.

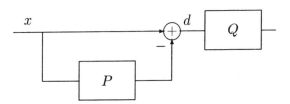

Figure 1.8: *D*PCM encoder. The decoder has the same structure as shown in Figure 1.6 b).*

Subband Coding

A subband encoder consists of two main parts:

- A critically sampled filter bank whose task it is to split the signal to be coded into subbands with almost no overlap in the frequency domain. The filtering task is performed by filters with unit sample responses $\{h_n(l)\}$, $n = 0, 1, \ldots N - 1$, see Figure 1.10. Assuming equal bandwidths for each of the N filters, as illustrated in Figure 1.9, it is reasonable to lower the sampling rate of the filter outputs by a factor of N since the bandwidth is one Nth of the original bandwidth. The subsampling operation is denoted $\downarrow N$ in Figure 1.10. This operation results in a baseband representation for all the subsampled subbands, see Appendix A. It is the combination of the collection of filters and subsamplers that constitute the critically sampled filter bank. Note that by critically subsampling the filter outputs, the total number of signal samples is the same before and after the filter bank.

- A set of different quantizers making it possible to represent the signals in each subband with as few bits as possible.

The subband decoder performs inverse operations of the two components described above. In particular the synthesis filter bank takes N subband signals as inputs. The sampling rate of the original signal, $x(l)$, is restored by the upsampling operation, denoted by $\uparrow N$ in Figure 1.10. Upsampling entails the insertion of $N - 1$ zero-valued samples between each incoming signal sample. Subsequently, appropriate filtering and summation operations are performed. In the absence of subband quantization the filter banks used are commonly designed for *perfect* or *almost perfect* reconstruction, i.e. $\tilde{x}(l) = x(l - l_0)$.

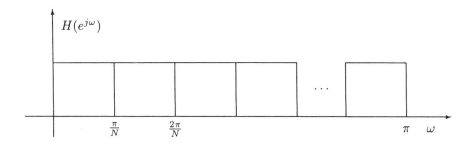

Figure 1.9: *Ideal "brick wall" frequency responses of the analysis channel filters in a subband coder using a* uniform *partitioning of the frequency band.*

The analysis filter bank attempts to *decorrelate* the signal. If the analysis filter bank succeeds in this endeavor, each subband sample carries almost separate information. The key to subjectively good compression lies in exploiting the different perceptual relevancy of the subbands. The more relevant or important subbands are given many bits for their representation, whereas the less relevant or less important subbands are given fewer bits.

The reconstructed signal can again be found directly from the decoder portion of Figure 1.10 by identifying the input to the upsampler in channel no. j as $e_j(l)$. This signal is the reconstructed subband signal. The expression for the output of the synthesis filter bank is given by [11]:

$$\tilde{x}(l) = \sum_{n=0}^{N-1} \sum_{k=-\infty}^{\infty} e_n(k)g_n(l - kN), \qquad (1.11)$$

Note again that this can be viewed as a series expansion of the signal $\tilde{x}(l)$. $\{g_n(l - kN), l = 0, 1, \ldots, N - 1, k \in \mathbf{Z}^1\}$ are basis functions representing the unit sample responses of the N subband reconstruction filters. The basis functions are used over and over again as for DPCM, but in SBC there are N basis functions which are shifted by N samples before they are used over again. This interpretation is more intuitively evident if we write Equation 1.11 in vector notation in the same way as was done in Section 1.3.1. In the present case we assume, for illustrative purposes, that

---

[1]$\mathbf{Z}$ denotes the set of integers.

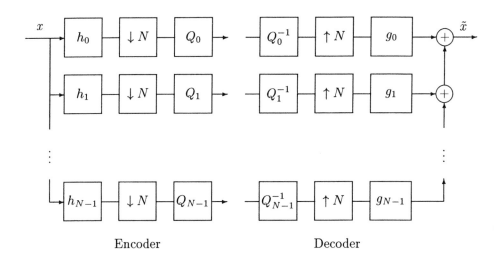

Figure 1.10: *Subband coder system.*

the number of channels, N, is equal to 2:

$$
\begin{bmatrix} \vdots \\ \tilde{x}(0) \\ \tilde{x}(1) \\ \tilde{x}(2) \\ \vdots \end{bmatrix} = \ldots + e_0(0) \begin{bmatrix} \vdots \\ g_0(0) \\ g_0(1) \\ g_0(2) \\ g_0(3) \\ \vdots \end{bmatrix} + e_0(1) \begin{bmatrix} \vdots \\ g_0(-2) \\ g_0(-1) \\ g_0(0) \\ g_0(1) \\ \vdots \end{bmatrix} +
$$

$$
\ldots + e_1(0) \begin{bmatrix} \vdots \\ g_1(0) \\ g_1(1) \\ g_1(2) \\ g_1(3) \\ \vdots \end{bmatrix} + e_1(1) \begin{bmatrix} \vdots \\ g_1(-2) \\ g_1(-1) \\ g_1(0) \\ g_1(1) \\ \vdots \end{bmatrix} + \ldots \quad (1.12)
$$

Figure 1.11 shows an example of four of the basis functions/vectors for a two-channel filter bank.

We conclude this section by pointing out that *transform coding* is just a special case of subband coding. This will be formally established in the next chapter.

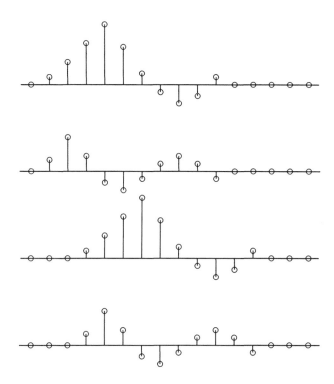

Figure 1.11: *Four basis functions of a two-channel synthesis filter bank. Notice that the two functions are repeated after a shift by two samples. The two functions are repeated infinitely in the same manner.*

Vector Quantization

Vector quantization (VQ) was introduced in Section 1.2.4 as the ultimate coding method. Here we consider a practical version of VQ which is block oriented just as transform coding. Unlike the other coding schemes presented so far, the signal reconstruction does not rely on a set of basis functions, but the reconstructed signal is an approximation of the original signal composed of a set of candidate functions taken from a codebook, as described in Section 1.2.4. Usually one function from the codebook is used in the approximate representation. Then the decoder is simply a look-up table to which an address is transmitted from the encoder indicating which word $g_j(l)$, of length N samples, is going to be output.

In *multistage VQ* the reconstruction of one block of data (consisting of

N samples) can be expressed as

$$\tilde{x}(l) = \sum_{k=0}^{K-1} e_k g_{j(k)}(l), \quad l \in 0, 1, \ldots, N - 1. \tag{1.13}$$

The code in multistage VQ consists of the scaling factors

$$\{e_k, \ k = 0, 1, \ldots, K - 1\}, \tag{1.14}$$

and the codebook indices

$$\{j(k), \ k = 0, 1, \ldots, K - 1\}. \tag{1.15}$$

The signal is composed of K scaled codebook vectors, each selected in the encoder from different, optimized codebooks. Although the structure also in this case looks like a series expansion, a study of the encoder reveals an even clearer distinction to the other coders.

Figure 1.12 illustrates the encoder structure. The first block in the en-

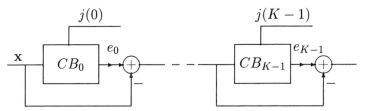

Figure 1.12: *Multistage VQ encoder structure showing the successive approximation of the signal vector.*

coder makes a rough approximation of the input vector by selecting the codebook vector which upon scaling by e_0 is closest in some distortion measure. Then this approximation is subtracted from the input signal. In the second stage the difference signal is approximated by a vector from the second codebook and scaled by e_1. This procedure continues in K stages. In the encoder we are thus making successive approximation to the input vector. The "basis functions" are chosen from a pool of possible candidates, and at each stage they are chosen from different codebooks. The composition of the codebooks is done by codebook optimization from representative training data, as will be discussed in Chapter 4.

A special case of this structure is the *mean-gain-shape VQ* [7], where the first function is a constant representing the mean, and the second function is selected from a codebook. The scaling factor for this vector is therefore the rms value of the block.

A possible interpretation of multistage VQ is that it uses adaptive approximate basis functions. However, the representation model is never complete in VQ. The number of vectors used in the approximation is always much smaller than the dimensionality of the problem. This is because VQ combines the representation and approximation stages. This will become clearer in Chapter 4.

In the next chapter the series expansion representation for frequency domain coders is the main topic.

Chapter 2

Exact Signal Decomposition

Signal decomposition is a way of changing the signal representation to a form more amenable to subsequent quantization/coding. The decomposition usually entails decorrelation of the signal such that each component in the new representation carries more or less separate information. In this book we will only discuss minimum representations, that is, the total number of samples in the original and decomposed representations are equal. By the term *representation* we mean a complete description of the original signal. This implies that the decomposition has an exact inverse.

As demonstrated in Chapter 1, signal decomposition can be viewed as a series expansion. In the present chapter we will elaborate on series expansions and show that a filter bank decomposition is nothing but a series expansion of the signal with strictly defined relations between the basis vectors. In this chapter, our main concern will be the invertibility of the decomposition. A signal decomposition with an exact inverse is said to possess the *perfect reconstruction* (PR) property. The design of *useful* filter banks with the PR property is treated in Chapters 3, 5, and 8.

2.1 Series Expansion: Block Transforms

In this section we review decomposition of finite length signals as a classical series expansion problem. This discussion will serve as the basis for more general signal decompositions to be treated in the next section.

In many applications (e.g. transform coding) the overall signal is split into blocks, each of which is decomposed individually. This is called *block decomposition.* The following discussion applies to such cases and, of course, to finite length signals where the whole signal is decomposed as one block.

Let $\mathbf{x} = [x(0), x(1), \ldots, x(N-1)]^T$ be a column vector representing the input signal. Assume that the signal components are real valued, i.e. $\mathbf{x} \in \mathbf{R}^N$. Let furthermore $\boldsymbol{\phi}_i = [\phi_i(0), \phi_i(1), \ldots, \phi_i(N-1)]^T$, $i = 0, 1, \ldots, N-1$, be an arbitrary set of N *linearly independent* complex vectors, i. e. $\boldsymbol{\phi}_i \in \mathbf{C}^N$. Linear independence means that no vector in the set can be expressed as a linear combination of the other vectors in the set. Furthermore, if the vectors are *orthogonal* they have the following property:

$$\boldsymbol{\phi}_i^H \boldsymbol{\phi}_j = c_i \delta_{ij} = \begin{cases} c_i & \text{for } i = j \\ 0 & \text{otherwise.} \end{cases} \tag{2.1}$$

$\{c_i, \ i = 0, 1, \ldots, N-1\}$ are constants, δ_{ij} is the Kronecker delta, and H denotes the *Hermitian* transpose, i.e. the complex conjugate of the transpose of a vector or matrix. For unit length orthogonal vectors $\boldsymbol{\phi}_i^H \boldsymbol{\phi}_j = \delta_{ij}$.

Assuming only linear independence for now, the vectors $\boldsymbol{\phi}_i \in \mathbf{C}^N$, $i = 0, 1, \ldots, N-1$, can be selected as a basis in $\mathbf{R}^N$ (or $\mathbf{C}^N$ if desired). Thus, $\mathbf{x}$ as defined above, can be written as

$$\mathbf{x} = \sum_{i=0}^{N-1} a_i \boldsymbol{\phi}_i = \boldsymbol{\Phi}\mathbf{a}, \tag{2.2}$$

where $\boldsymbol{\Phi}$ is the $N \times N$ matrix with the basis vectors as columns,

$$\boldsymbol{\Phi} = [\boldsymbol{\phi}_0, \boldsymbol{\phi}_1, \ldots, \boldsymbol{\phi}_{N-1}], \tag{2.3}$$

and $\mathbf{a}$ is the column vector of expansion coefficients

$$\mathbf{a} = [a_0, a_1, \ldots, a_{N-1}]^T. \tag{2.4}$$

Writing the above matrix relation out in more detail we have

$$\begin{bmatrix} x(0) \\ x(1) \\ \vdots \\ x(N-1) \end{bmatrix} = \begin{bmatrix} \phi_0(0) & \phi_1(0) & \cdots & \phi_{N-1}(0) \\ \phi_0(1) & \phi_1(1) & \cdots & \phi_{N-1}(1) \\ \vdots & \vdots & & \vdots \\ \phi_0(N-1) & \phi_1(N-1) & \cdots & \phi_{N-1}(N-1) \end{bmatrix} \begin{bmatrix} a_0 \\ a_1 \\ \vdots \\ a_{N-1} \end{bmatrix}.$$

For each signal sample, $x(l)$, we can consequently write

$$x(l) = \sum_{i=0}^{N-1} a_i \phi_i(l). \tag{2.5}$$

Given the linear independence of the columns of $\boldsymbol{\Phi}$, it is evident that the vector of expansion coefficients can be obtained from

$$a = \boldsymbol{\Phi}^{-1}x. \tag{2.6}$$

Introducing the matrix $\boldsymbol{\Psi} = (\boldsymbol{\Phi}^{-1})^H$, a can be expressed as

$$a = \boldsymbol{\Psi}^H x. \tag{2.7}$$

The columns of $\boldsymbol{\Psi}$ are called the *reciprocal* basis vectors [12]. Identifying these reciprocal basis vectors as $\boldsymbol{\psi}_i$, $i = 0, 1, \ldots, N-1$, we have that

$$\phi_i^H \psi_j = \delta_{ij} = \begin{cases} 1 & \text{for } i = j \\ 0 & \text{otherwise.} \end{cases} \tag{2.8}$$

Each individual expansion coefficient can be obtained through

$$a_i = \boldsymbol{\psi}_i^H x, \ i = 0, 1, \ldots, N-1. \tag{2.9}$$

In conclusion, the general expression for the series expansion is

$$x(l) = \sum_{i=0}^{N-1} (\boldsymbol{\psi}_i^H x)\phi_i(l). \tag{2.10}$$

2.1.1 Unitary Transforms

For the particular and important case when the basis is chosen to be *orthonormal* we have

$$\boldsymbol{\Phi}^H = \boldsymbol{\Phi}^{-1}, \tag{2.11}$$

and consequently

$$\boldsymbol{\Psi} = \boldsymbol{\Phi}. \tag{2.12}$$

That is, the matrix of reciprocal basis vectors is equal to the matrix of original basis vectors. A matrix with the property given in Equation 2.11 is said to be *unitary*.

For the unitary case the coefficient vector can be found from

$$a = \boldsymbol{\Phi}^H x, \tag{2.13}$$

and the individual coefficients in Equation 2.5 can be computed as

$$a_i = \phi_i^H x, \ i = 0, 1, \ldots, N-1. \tag{2.14}$$

In the unitary case the expression for the series expansion is

$$x(l) = \sum_{i=0}^{N-1} (\phi_i^H \mathbf{x}) \phi_i(l). \tag{2.15}$$

Example: The discrete Fourier transform (DFT)

In a unitary version of the DFT, the transform matrix elements are given by

$$\psi_k(n) = \frac{1}{\sqrt{N}} e^{-j\frac{2\pi}{N}kn}, \tag{2.16}$$

where $j = \sqrt{-1}$. Since the transform is unitary, the inverse transform is simply the Hermitian transpose of the forward transform matrix. As the transform in this case is symmetric, there is no need for transposition of the matrix. We are then left only with the complex conjugation of the transform matrix to obtain the inverse matrix:

$$\phi_n(k) = \frac{1}{\sqrt{N}} e^{j\frac{2\pi}{N}kn}. \tag{2.17}$$

In the signal processing literature it is not common to use the unitary version of the DFT matrix. The scaling factor $1/\sqrt{N}$ in Equation 2.16 is then left out. This implies that the inverse matrix elements of Equation 2.17 must be scaled by $1/N$ rather than $1/\sqrt{N}$.

□

2.2   Series Expansion: Multirate Filter Banks

The previous discussion is closely related to block oriented *transform coding*. We shall now generalize the series expansion approach to encompass all basic frequency domain decompositions.

Let $x(l)$ be an infinite length finite energy sequence, i.e.

$$\sum_{l=-\infty}^{\infty} x^2(l) < \infty. \tag{2.18}$$

It is customary to call such a sequence an ℓ^2-sequence. Any practical signal will be finite such that the ℓ^2 requirement is fulfilled. For intra-frame image coding this is obvious.

The most common procedure in signal decomposition is to map the input vector **x** onto a basis, that is, make a series expansion, which for finite length signals, has a number of terms equal to the number of signal components. A series expansion in ℓ^2 is a generalization of Equation 2.5 to infinite length summation

$$x(l) = \sum_{i=-\infty}^{\infty} a_i \phi_i(l).$$ \hfill (2.19)

In this equation $\{a_i, i \in \mathbf{Z}\}$ are the coefficients and $\{\phi_i(l), i \in \mathbf{Z}\}$ is a basis in ℓ^2. We can still write this equation as a matrix representing the basis vectors multiplied by a coefficient vector as in Equation 2.2, but in this case the matrix and vectors have infinite dimensions. Although the signal is assumed to be of infinite extent, the basis functions may have finite, even quite limited regions of support as long as we have an infinite number of them.

To make practical, implementable systems we will impose certain restrictions on the basis vectors. The most fundamental restriction is to introduce shift relations between the basis vectors. Let N vectors be a *generating set* $\{g_i(l)\} = \{\phi_i(l), i = 0, 1, \ldots, N-1, l \in \mathbf{Z}\}$, and generate new vectors through shifts of the generating set according to

$$\phi_i(l) = g_{((i))_N}(l - \lfloor \tfrac{i}{N} \rfloor N), \ l \in \mathbf{Z}, \ i \in \mathbf{Z}$$ \hfill (2.20)

where $((k))_N$ is k *modulo* N and $\lfloor k \rfloor$ is the largest integer smaller or equal to k. If all these vectors are linearly independent, they may form a basis for infinite dimensional signals.

Example: Generating basis vectors for $N = 2$.

In this case we have a generating set $g_0(l)$ and $g_1(l)$, and an infinite basis is formed according to Equation 2.20, as

$$\vdots$$

$$
\begin{aligned}
\phi_{-4}(l) &= g_0(l+4) \\
\phi_{-3}(l) &= g_1(l+4) \\
\phi_{-2}(l) &= g_0(l+2) \\
\phi_{-1}(l) &= g_1(l+2)
\end{aligned}
$$

$$
\begin{aligned}
\phi_0(l) &= g_0(l) \\
\phi_1(l) &= g_1(l) \\
\phi_2(l) &= g_0(l-2) \\
\phi_3(l) &= g_1(l-2) \\
\phi_4(l) &= g_0(l-4) \\
\phi_5(l) &= g_1(l-4) \\
&\vdots
\end{aligned}
$$

$\square$

A generating set of N basis vectors of length L (L can possibly be infinite) can be collected in the $L \times N$ matrix

$$
\mathbf{G} = [\mathbf{g}_0, \mathbf{g}_1, \ldots, \mathbf{g}_{N-1}], \tag{2.21}
$$

where the vectors $\{\mathbf{g}_i, \ i = 0, 1, \ldots, N-1\}$ are column vectors. With this definition of the generating matrix, the complete system of basis vectors is given by the $\mathbf{\Phi}$-matrix which is illustrated in Figure 2.2 for basis vectors of length $L > N$. The shift relation between the basis vectors is seen through the vertical displacement between neighboring blocks representing the submatrix $\mathbf{G}$.

With the basis vectors of Equation 2.20, the series expansion can be written as

$$
x(l) = \sum_{i=-\infty}^{\infty} a_i g_{((i))_N} \left(l - \lfloor \frac{i}{N} \rfloor N\right). \tag{2.22}
$$

To obtain a filter bank implementation of this series expansion, let us split the summation into a double summation by replacing the summation index i by n and k according to $i = n + kN$, $n = 0, 1, \ldots, N-1$, $k \in \mathbf{Z}$. We then obtain

$$
x(l) = \sum_{n=0}^{N-1} \sum_{k=-\infty}^{\infty} g_n(l - kN) y_n(k), \tag{2.23}
$$

where we have, for convenience, introduced $y_n(k) = a_{n+kN}$. In the next section we will see that $y_n(k)$ can be interpreted as the subband signal in channel n. One may also sometimes need the alternative convolution expression which is obtained by the substitution $k = \lfloor l/N \rfloor - m$:

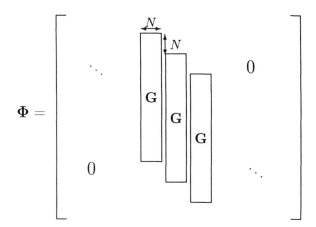

Figure 2.1: *Illustration of the synthesis matrix where the column vectors are the basis vectors. Each block denoted* **G** *contains the N basis vectors of the generating set. The neighboring generating set is shifted vertically by N samples.*

$$x(l) = \sum_{n=0}^{N-1} \sum_{m=-\infty}^{\infty} g_n(mN + ((l))_N)y_n(\lfloor l/N \rfloor - m). \qquad (2.24)$$

The expression $((l))_N$ runs over the N different phases in the interpolation process (see Appendix A) and selects the appropriate set of vector samples (coefficients). Note that the summation over m is finite or semi-infinite when the filters are either of the FIR type or causal IIR type.

Using vector notation $\mathbf{x} = \mathbf{\Phi}\mathbf{y}$, the $\mathbf{y}$-vector is given by

$$\mathbf{y} = [\ldots, y_0(n-1), y_1(n-1), \ldots, y_{N-1}(n-1), y_0(n), y_1(n), \ldots, y_{N-1}(n), \ldots]^T. \qquad (2.25)$$

If we choose the length of the basis vectors equal to N, the number of basis vectors in the generating set, then the expansion corresponds to the block transforms of the previous section. For that special case the outer summation in Equation 2.23 represents the decomposition of a block, and the inner summation index identifies the blocks. It is therefore more meaningful to interchange the order of summations in this case as

$$x(l) = \sum_{k=-\infty}^{\infty} \left[\sum_{n=0}^{N-1} y_n(k) g_n(l - kN) \right]. \qquad (2.26)$$

The operation in the bracket is a square transform: It transforms a vector consisting of one sample from each channel $\{y_n(k),\ n = 0, 1, \ldots, N - 1\}$ to N output samples of $x(l)$. There are no overlapping contributions to the output signal from different blocks. The matrix for the transformation of the whole signal is composed of square $N \times N$ transforms G as shown in Figure 2.2.

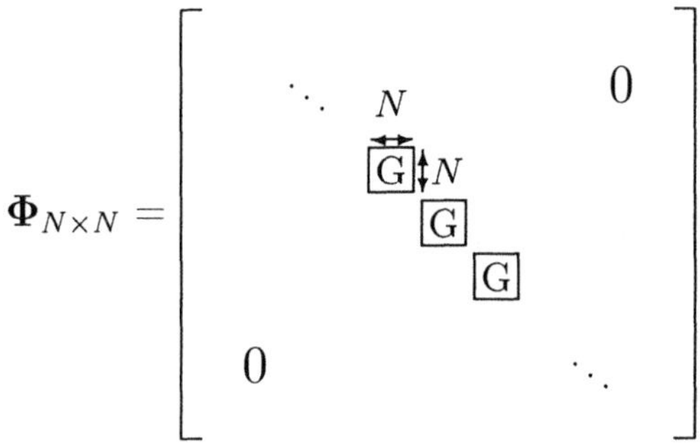

Figure 2.2: *Matrix structure of a system consisting of submatrices G of size $N \times N$. The submatrices contain the generating set of basis vectors. Each block represents the transform in transform coding systems.*

2.2.1   Synthesis Filter Bank Interpretation

The reason why we introduced the shift relationship between the basis functions was to make it possible to interpret, and implement, Equation 2.19 as a parallel filter bank described by Equations 2.23 and 2.24. The outer summation indicates that the output signal $x(l)$ of the system is composed of N contributions. We denote each contribution a *channel* or a *subband*. The inner summation in Equation 2.23 represents time discrete convolution in channel no. n between the input signal $y_n(k)$ and a filter with unit sample response $g_n(k)$. The factor N in the argument of g_n in Equation 2.23 accounts for up-sampling by a factor N. In this way filter no. n produces N times as many output samples as there are samples in $\{y_n(k),\ k \in \mathbf{Z}\}$. By doing so, the summation of each of the N contributions in Equation 2.23 results in the same number of samples as the total number of subband samples. In Figure 2.3 we have illustrated the synthesis filter bank interpretation of Equation 2.23.

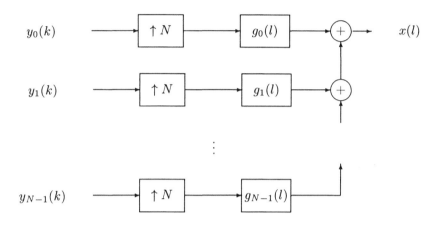

Figure 2.3: *Synthesis multirate filter interpretation of the series expansion of an infinite dimensional signal.*

2.2.2 Inversion Formula: Analysis Filter Bank

We conjecture that the coefficients in the infinite dimensional series expansion can also be found from the reciprocal basis as in Equation 2.9. A proof will be given in Section 2.3.

If using a notation analogous to that of Section 2.1, where the matrix $\mathbf{\Phi}$ represents the basis, and the matrix $\mathbf{\Psi}$ represents the reciprocal basis, the coefficient vector $\mathbf{a}$ can be expressed as

$$\mathbf{a} = \mathbf{\Psi}^H \mathbf{x}. \tag{2.27}$$

At this point we do not know whether the vectors in the reciprocal basis have the same shift relationship as the vectors in the original basis. This is a necessary property for the implementation of the analysis system as a filter bank. In the unitary case, this is obvious, because the matrix representing the reciprocal basis is the transpose of the matrix representing the original basis.

Let us consider the general case. The reciprocal basis must have basis vectors that are orthogonal to all but one of the vectors in the basis according to

$$\phi_i^H \psi_j = \delta_{ij} = \begin{cases} 1 & \text{for } i = j \\ 0 & \text{otherwise.} \end{cases} \tag{2.28}$$

Recall that the matrix $\mathbf{G}$ is a collection of the N generating vectors of the basis and that all the other vectors in the basis are shifted versions of the generating set. Assume that there exist N vectors in the reciprocal basis, where each vector is orthogonal to all vectors, except for one, in the matrix of original basis vectors, $\mathbf{\Phi}$. We will call them the *generating set for the reciprocal basis* and collect them in the matrix

$$\mathbf{H} = \left[\tilde{\mathbf{h}}_0, \tilde{\mathbf{h}}_1, \ldots, \tilde{\mathbf{h}}_{N-1} \right], \tag{2.29}$$

where each $\tilde{\mathbf{h}}_k$ is a column vector. These vectors are related to the unit sample responses of an analysis filter bank, as will be shown shortly.

If the matrix of generating vectors $\mathbf{H}$ has been found, we may generate other sets of N vectors composed of shifted (by kN, $k \in \mathbf{Z}$) versions of the generating set. These vectors are also orthogonal to the columns of $\mathbf{\Phi}$ due to the periodicity of the basis vectors (see Figure 2.2). The new vectors will therefore automatically be true members of the reciprocal basis. This can be done for any integer value of k providing the infinite number of necessary vectors in the reciprocal basis $\mathbf{\Psi}$.

We may sum up the obtained results as follows: Given a basis $\mathbf{\Phi}$ in ℓ^2 with the required filter bank property that a generating set of N basis vectors are used to generate all the others by integer shifts of kN samples, this representing the synthesis filter bank, a necessary and sufficient condition for finding the expansion coefficients (subband signals), is the existence of the reciprocal basis with the property given in Equation 2.28. The reciprocal basis has the same shift property as the basis and can therefore be implemented as a filter bank. Figure 2.4 shows the $\mathbf{\Psi}$-matrix structure.

We are now able to formulate the analysis filter bank equation. Each

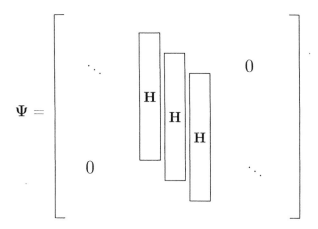

Figure 2.4: *Illustration of the analysis filter bank matrix. Each block denoted* **H** *represents the generating set.*

element of **a** of Equation 2.27 can be written as

$$a_l = \sum_{i=-\infty}^{\infty} \psi_l^*(i)x(i), \tag{2.30}$$

where $\psi_l^*(i)$ is element no. i in row no. l of the matrix $\boldsymbol{\Psi}^H$. The shift property between the reciprocal basis vectors can be expressed as

$$\psi_{n+kN}^*(i) = \psi_n^*(i - kN) = \tilde{h}_n^*(i - kN), \ n = 0, 1, \ldots, N - 1, \ k \in \mathbf{Z}, \tag{2.31}$$

where $\tilde{h}_n(i)$ are the components of one vector in the generating set in the reciprocal basis. If we replace l by $n + kN$ in Equation 2.30, we get

$$a_l = a_{n+kN} = y_n(k) = \sum_{i=-\infty}^{\infty} \tilde{h}_n^*(i - kN)x(i). \tag{2.32}$$

We generate all values of l by constraining n to the set $n \in 0, 1, \ldots, N - 1$ while k may be any integer. We observe that all coefficients $y_n(\cdot)$ are found by cross-correlating the input signal and one basis vector from the generating set. Only every Nth of the possible cross-correlation lags are calculated.

It is easy to replace the cross-correlation operation by convolution. To obtain that, we introduce unit sample responses $h_n(i) = \tilde{h}_n^*(L - 1 - i)$

for length L basis vectors, leading to a convolution based version of Equation 2.32:

$$y_n(k) = \sum_{i=-\infty}^{\infty} h_n(kN - i)x(i). \qquad (2.33)$$

The matrix $\mathbf{H}$ can then be expressed by the unit sample responses as

$$\mathbf{H} = \begin{bmatrix} h_0^*(L-1) & h_1^*(L-1) & \cdots & h_{N-1}^*(L-1) \\ h_0^*(L-2) & h_1^*(L-2) & \cdots & h_{N-1}^*(L-2) \\ \vdots & \vdots & \cdots & \vdots \\ h_0^*(0) & h_1^*(0) & \cdots & h_{N-1}^*(0) \end{bmatrix}. \qquad (2.34)$$

Equation 2.32 can be interpreted as representing a *parallel analysis filter bank*. The input signal $x(l)$ is run through N parallel filters with unit sample response $h_n(l)$ in channel no. n. The output signals are subsampled by a factor of N and represented by $\{y_n(l),\ n = 0, 1, \ldots, N-1\}$. The analysis filter bank is illustrated in Figure 2.5.

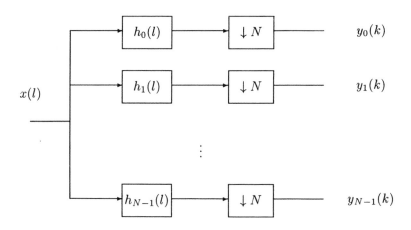

Figure 2.5: *Analysis filter bank structure.*

Although detailed filter bank structures with potentially perfect reconstruction properties are the main topic for the next chapter, we make some general comments at this point. As pointed out earlier, the reciprocal basis is equal to the basis in the case of unitary systems (when all the basis vectors are mutually orthonormal). Then if one basis contains finite length functions, so will the other. In general, there is no guarantee that this will be true. As we will see in the next chapter, the synthesis filter bank may contain infinite length filters even when the analysis filters are all of finite length and vice versa. It may therefore appear as if the only reasonable filter banks which can be constructed are the unitary ones. Even then it is not a trivial problem to design good filter banks. In fact, most of the transform and subband coder literature considers unitary filter banks only. Considering the volume of that literature, it is easily realized that the design task is considerable also for the unitary case. However, in Chapters 5 and 8 we are going to argue that the nonunitary case offers more flexibility in the filter bank design and thus can provide better subband coders.

The intention of this section has been to show the matrix formulation for signal decomposition and present the relation between the analysis and synthesis matrices. As a special case, we have chosen the synthesis matrix of a particular form to make it implementable as a critically sampled synthesis filter bank. Then a set of N columns of the matrix $\mathbf{\Psi}$ constitutes reversed and complex conjugated versions of the unit sample responses in an analysis filter bank of N uniformly spaced channels. The rest of the matrix is composed of horizontally shifted (by kN, $k \in \mathbf{Z}$) samples of this set. The synthesis filter bank has a structure like the network transpose [11] of the analysis filter bank. For the particular case of a unitary system, the unit sample responses of the analysis and synthesis filters are related through reversion and complex conjugation.

Arbitrary decompositions, including adaptive filter systems, can be described by the matrix formulation. This system description is therefore general.

2.3   The PR Conditions and Biorthogonality

We will now give an alternative derivation of the PR conditions for the special case of signal decomposition by multirate filter banks. In this derivation we avoid inversion of infinite-dimensional matrices describing filtering of infinite-length signals.

Consider a uniform analysis filter bank followed by a synthesis filter bank which together constitute a perfect reconstruction system. If cascading another analysis filter bank at the output of the synthesis filter bank, the same output will again be produced as in the first case. It follows that the cascade

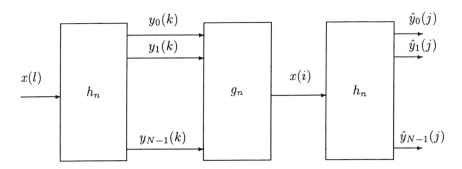

Figure 2.6: *Back-to-back configuration of synthesis and analysis filter banks.*

of a synthesis filter bank with an analysis filter bank also forms a perfect reconstruction system if the reverse order gives perfect reconstruction. The system is illustrated in Figure 2.6. The unit sample response in channel n in the synthesis filter bank is again called $g_n(l)$ and the corresponding response in the analysis filter $h_n(l)$. The contribution to $x(l)$ from channel n is called $x_n(l)$, and is found through discrete convolution including upsampling, as in Equation 2.23,

$$x_n(l) = \sum_{k=-\infty}^{\infty} g_n(l - kN)y_n(k).$$ (2.35)

The complete synthesized signal is given as the sum of the different contributions,

$$x(l) = \sum_{n=0}^{N-1} x_n(l).$$ (2.36)

The jth output from the analysis filter in channel m is calculated using Equation 2.33[1].

$$\hat{y}_m(j) = \sum_{l=-\infty}^{\infty} h_m(jN - l)x(l).$$ (2.37)

Combining equations 2.35, 2.36, and 2.37 and interchanging the order of

---

[1]We drop complex conjugation in the notation from now on since in many instances we assume the filters to be real.

summations, we get

$$\hat{y}_m(j) = \sum_{n=0}^{N-1} \sum_{k=-\infty}^{\infty} y_n(k) \sum_{l=-\infty}^{\infty} g_n(l-kN)h_m(jN-l). \tag{2.38}$$

To obtain perfect reconstruction in the sense that the output signals are delayed versions of the input signals, that is $\hat{y}_m(j) = y_m(j - r_m)$, the only possible solution is

$$\sum_{l=-\infty}^{\infty} g_n(l-kN)h_m((j-r_m)N-l) = \delta_{nm}\delta_{k(j-r_m)}. \tag{2.39}$$

Notice that we have allowed for different delays in the different channels. The perfect reconstruction condition in Equation 2.39 shows that the analysis filter response in channel no. m must be orthonormal to all reversed and shifted by $pN + r_m$ ($p \in \mathbf{Z}$) versions of the synthesis responses except for the original version of g_n. This is usually referred to as *biorthogonality*. This is, of course, the same result as derived from the properties of the reciprocal basis. In the previous section, however, we did not explicitly mention the possibility of using different delays in the different channels.

2.4 Extension of Finite Length Signals

So far in this chapter we have shown that an analysis-synthesis filter bank system can be designed to preserve the perfect reconstruction property in the absence of quantization noise.

For transform coding, utilizing nonoverlapping transforms, reconstruction of finite length signals poses no problems, since the input signal is simply partitioned into equal-sized blocks, and perfect reconstruction is guaranteed for each block. In contrast, for proper subband coding nontrivial problems arise when filtering finite length signals. This is due to the overlapping unit pulse responses of the analysis and synthesis filter channels: A signal segment will only be properly reconstructed with the added influence of adjacent signal parts.

Therefore, we have to pay special attention to the filtering of the two extremities of the signal. A simple solution would be to extend the signal with a known, deterministic signal, for example by replicating the endpoint value as done by Karlsson and Vetterli [13], but this invariably leads to more

subband samples to code and transmit to the decoder. Another suggested
method in the case of IIR filter banks is to send as side information the
memory content in each channel filter as the signal end is reached. This is
also an insatisfactory solution due to the increase in bit rate.

There are only two known methods of extending the finite length input
signal while preserving the perfect reconstruction property and not generating side information: The *circular extension* and the *mirror extension*.
Both methods extend the finite length signal so as to make it periodic –
and therefore also infinite length – thus securing the perfect reconstruction
property guaranteed for all infinite length signals.

2.4.1   The Circular Extension Method

Figure 2.7 shows the circular extension of a finite length ramp signal. If the
original signal has length K samples, the extended signal is a periodic signal
with period K. It can easily be proven that time-invariant, linear filtering
preserves the periodic property of an input signal. Therefore, each channel
signal will, before decimation, be periodic with period K. Assume that

$$K = pN, \qquad (2.40)$$

where N is the decimation factor and p is a positive integer. Each of the
decimated subband signals is now periodic with period p.

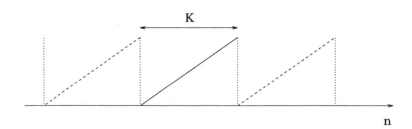

Figure 2.7: *Circular extension of ramp signal.*

For perfect reconstruction of the original signal, it is necessary that the
decoder has full knowledge of each infinite length subband signal. But due
to the fact that they are all periodic with period p, it suffices to transmit

only p samples from each band. The subband samples outside this range are easily found by circular extension.

Since circular extension and filtering is the time equivalent of frequency-domain *discrete Fourier transform* (DFT) operations, it is also possible to prove the perfect reconstruction property using DFTs instead of *discrete-time Fourier transforms* (DTFTs) when deriving filter bank expressions.

The advantages of the circular extension method are: Simplicity, both linear and nonlinear phase filters can be used, and implementations can utilize *fast Fourier transform* (FFT) algorithms. The major drawback is that the circular extension tends to produce artificial steps in the original signal. This is evident from the ramp example in Figure 2.7. In a graytone image coding context this means that any image where the left/right or top/bottom image regions have different graytone values, the analysis filter bank "perceives" an abrupt signal transition. Fourier relations tell us that this type of signal contains high frequencies. This leads to significant signal energy in the higher subbands, which again degrades compression performance because this artificially generated information must be coded and transmitted. At low bit rates, the circular extension method may generate ringing noise artifacts along the image borders. This happens when the high frequency channels are inadequately represented.

2.4.2 The Mirror Extension Method

As for the circular extension method, this method preserves the perfect reconstruction property by extending the original, finite length signal into an infinite length periodic signal. The difference lies in the extension: Instead of simply repeating the signal on the left and right, the signal is now extended by first performing a mirror reflection at one endpoint and thereafter making periodic extensions of the double length signal. This method was introduced by Smith and Eddins [14]. The procedure is shown in Figure 2.8.

The rationale for using this type of extension is that it avoids generating signal discontinuities like those of the circular extension. However, preserving the number of signal samples to be transmitted is more tricky in the mirror extension case because the extended signal now has period $2K$. Rather than exploiting periodicity, the mirror extension rely on *preserving the symmetry across both mirror points*. If a symmetric signal $x(n)$ is filtered with a linear phase filter $h(n)$, the output signal will be symmetric according to the rules given in Table 2.1 [15]. Half-sample symmetry refers

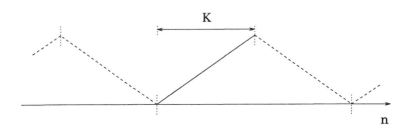

Figure 2.8: *Mirror extension of ramp signal.*

to sequences where the point of symmetry is located between two samples.

$x(n)$	$h(n)$	$(x * h)(n)$
Whole-sample symmetry	Whole-sample symmetry	Whole-sample symmetry
Half-sample symmetry	Half-sample symmetry	Whole-sample symmetry
Half-sample symmetry	Whole-sample symmetry	Half-sample symmetry
Symmetric	Symmetric	Symmetric
Anti-symmetric	Anti-symmetric	Symmetric
Symmetric	Anti-symmetric	Anti-symmetric

Table 2.1: *Symmetry relations for the convolution operator.*

Example: Mirror extension of even-length signal.

Consider an input signal of length 16 which is to be filtered with an even-length symmetric filter, $\mathbf{h} =$(-0.25, 0.75, 0.75, -0.25). The filter has half-sample symmetry. When extending the input signal we can choose whether to use whole- or half-sample symmetry. From Table 2.1 we observe that the convolution of two half-sample symmetric sequences yields a sequence with whole-sample symmetry. The trick is now to *extend the input signal such that the filtered signal will have whole-sample symmetry.* We therefore use half-sample symmetric extension of the input signal. Figure 2.9 depicts the extended and filtered signal prior to decimation. The dotted lines indicate the mirrored versions of the original signal.

Note that our filtered signal now has 17 distinct sample values. Critical decimation with factor N means that we are only allowed to transmit $16/N$

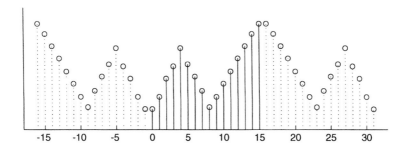

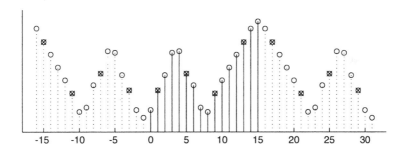

Figure 2.9: *Extended (top) and filtered (bottom) input signal.*

out of 17 samples. This is achieved by fulfilling two criteria:

1. Avoid picking the whole-sample symmetry samples, i.e. -1 and 17.

2. Make sure that the corresponding samples on the other side of the symmetry points have the correct distance. For example, when $N = 4$, we can not start with sample number 0, 4, etc. because then it will be necessary also to know sample number -4, -8, etc. If we start with sample number 1, 5, etc., the corresponding negative sample numbers are -3, -7, etc., and *they have identical values*.

The reader can confirm that these are the only possible choices:

$N = 2$: (0, 2, 4, 6, 8, 10, 12, 14).

$N = 4$: (1, 5, 9, 13).

$N = 8$: (3,11).

For $N = 4$ the correct choice of samples are indicated in Figure 2.9 with ×'es.

$\square$

The example above dealt with the even-length symmetric filter case. Martucci [15] gives a thorough derivation of all cases for a two-channel filter bank system.

Unfortunately, mirror extension is not possible for filters having nonlinear phase responses, since the method rests on the preservation of symmetry across the mirror reflection points. Therefore, the circular method must be used in the IIR filter bank case. In Chapter 7 we demonstrate the benefits of the mirror extension through an image coding example.

2.5   A Design Method for FIR Filter Banks

In the previous sections we have established that *biorthogonality* is the most general PR condition for critically sampled, uniform, multirate filter banks. The mutual orthogonality between analysis and synthesis basis vectors does not provide any relation between the length of the basis vectors of the analysis and the synthesis sides. For the special case of unitary analysis systems, the synthesis basis vectors are of the same length. In all other cases we have to impose additional constraints upon the basis vectors to make equal length vectors, if that is desirable. As we shall demonstrate in the next chapter, it is possible to find *filter structures* that yield PR and equal length analysis and synthesis vectors. For now, we will consider a parallel, unconstrained system where we derive design equations for the case of equal length vectors.

The design of perfect, or almost perfect reconstruction FIR filter banks, has received considerable attention and several solutions have been suggested [16, 17, 18]. In the uniform, N-channel, parallel case, Vetterli [19] showed that designing analysis/synthesis FIR systems with analysis filter lengths L means satisfying $L - N + 1$ nonlinear equations by choosing LN analysis filter coefficients. He also proposed a method for reducing this problem into a linear one. Unfortunately, this approach has a serious drawback since no guarantee is given as to the length of the resulting synthesis unit pulse responses. In the worst case, they may be as large as $(L-1)(N-1)+1$. In this scheme it is also difficult to control the optimality (to be discussed in Chapters 5 and 8) of the total analysis/synthesis system.

A different approach was developed by Nayebi et al. [20, 21]. They use time domain conditions combined with a gradient search method to design perfect reconstruction, parallel FIR filter banks. The drawback is that the resulting filter bank solution may be suboptimal for a given error function. In the following subsections a different, more intuitive derivation of the main equation of the algorithm is given [22].

The Unitary Case

Conditions for perfect reconstruction are very simple when applying a unitary filter bank. In accordance with Equation 2.34 we denote by $\mathbf{H}$ the $L \times N$ matrix for which the nth column comprise the coefficients of the nth analysis filter channel.

Now assume that the incoming discrete-time signal $x(k)$ is of infinite length. To reconstruct the original signal $x(k)$ from $y(k)$ a necessary and sufficient condition is that the (infinite) matrix $\mathbf{\Psi}^H$ of Figure 2.4 should be invertible. In practice, our (image) signals will be of finite length, and special care must be taken at the signal boundaries. Section 2.4 showed extension techniques for this purpose, and in this section we concentrate on the necessary conditions for perfect reconstruction in the center part of the signal where possible boundary effects can be ignored.

Writing the transformed signal or subband signal $y(l)$ as

$$\mathbf{y} = \mathbf{\Psi}^H \mathbf{x}, \tag{2.41}$$

and restricting the transform matrix to be unitary, the inverse transform is simply the matrix Hermitian and the original signal is exactly reconstructed by

$$\mathbf{x} = \mathbf{\Psi} \mathbf{y} = \mathbf{\Phi} \mathbf{y}. \tag{2.42}$$

In principle, it is easy to find an orthogonal set of basis vectors. According to the Gram-Schmidt method [23] we know that we can generate a set of orthonormal basis vectors from any linearly independent set of basis vectors. However, we are requiring that all basis vectors are orthogonal, meaning that also shifted versions of a vector by $\{kN, \ k \in \mathbf{Z}\}$ samples must be orthogonal to all the unshifted vectors. This makes the problem more involved. An additional problem is to find vector sets that will generate meaningful filters.

The Nonunitary Case

A perfect reconstruction, nonunitary, FIR, parallel filter bank system is obtained by relaxing the orthogonality restrictions from the previous paragraph. This means that the basis vectors, or their shifted versions, do not have to be mutually orthogonal. This applies to both analysis and synthesis vectors, i.e. $\{\mathbf{h}_i\}$ and $\{\mathbf{g}_i\}$.

To reconstruct the original signal, the transform $\mathbf{\Psi}^H = \mathbf{\Phi}^{-1}$ derived from the reciprocal basis is applied. The inverse transform (if it exists) may have infinite length (IIR) basis vectors [19, 24]. However, let us assume that a perfect reconstruction inverse transform does exist for a given forward transform. Furthermore, assume that the inverse transform has a block structure similar to that of the forward transform, with maximum length $L = pN$ of the basis vectors, where p is a positive integer. As in Equation 2.21, denote by $\mathbf{G}$ the $L \times N$ submatrix with columns composed of the coefficients of the synthesis FIR filter channels[2].

For perfect reconstruction we require

$$\mathbf{\Phi}\mathbf{\Phi}^{-1} = \mathbf{\Phi}\mathbf{\Psi}^H = \mathbf{\Psi}^H\mathbf{\Phi} = \mathbf{I}. \tag{2.43}$$

In the above operations the submatrices $\mathbf{H}$ and $\mathbf{G}$ are multiplied under different mutual shifts. For a certain combination of alignment we obtain the pure, i.e. unshifted, product between the submatrices,

$$\mathbf{H}^T\mathbf{G} = \mathbf{I}_N, \tag{2.44}$$

where $\mathbf{I}_N$ is the identity matrix of size $N \times N$. For all other combinations, the submatrix products should be equal to the zero matrix.

To put the above statements in mathematical form, we split the matrices $\mathbf{H}$ and $\mathbf{G}$ into p submatrices $\mathbf{H}_i$ and $\mathbf{G}_i$, $i = 1, \ldots, p$, with dimension $N \times N$:

$$\mathbf{H}^T = [\mathbf{H}_1 | \mathbf{H}_2 | \ldots | \mathbf{H}_p] \tag{2.45}$$

$$\mathbf{G}^T = [\mathbf{G}_1 | \mathbf{G}_2 | \ldots | \mathbf{G}_p]. \tag{2.46}$$

---

[2]At this point it may be useful to review the structure of the matrices $\mathbf{\Psi}$ and $\mathbf{\Phi}$ as visualized in Figures 2.2 and 2.4.

Then the PR condition is given by

$$
\underbrace{\begin{bmatrix}
\mathbf{G}_p & 0 & 0 & \cdots & 0 \\
\mathbf{G}_{p-1} & \mathbf{G}_p & 0 & \vdots & 0 \\
\vdots & \vdots & \ddots & & \\
\mathbf{G}_1 & \mathbf{G}_2 & \mathbf{G}_3 & \cdots & \mathbf{G}_p \\
0 & \mathbf{G}_1 & \mathbf{G}_2 & \cdots & \mathbf{G}_{p-1} \\
\vdots & \vdots & \vdots & \ddots & \vdots \\
0 & 0 & 0 & \cdots & \mathbf{G}_1
\end{bmatrix}}_{\mathbf{A}}
\underbrace{\begin{bmatrix}
\mathbf{H}_1^T \\
\mathbf{H}_2^T \\
\mathbf{H}_3^T \\
\vdots \\
\mathbf{H}_p^T
\end{bmatrix}}_{\mathbf{H}}
=
\underbrace{\begin{bmatrix}
0 \\
\vdots \\
0 \\
\mathbf{I}_N \\
0 \\
\vdots \\
0
\end{bmatrix}}_{\mathbf{B}},
\tag{2.47}
$$

where the position of the $N \times N$ identity matrix $\mathbf{I}_N$ is at the center of the $\mathbf{B}$ matrix. This is a matrix formulation of Equation 2.39 for the case of FIR filters. For IIR filters the matrix in Equation 2.47 is infinite-dimensional, and therefore not useful for system optimization.

To summarize, Equation 2.47 is equivalent to having a perfect reconstruction, FIR, uniform filter bank, with filter response lengths limited by $L = pN$.

The Nayebi Method

In [20, 21] Nayebi et al. derived an equation similar to Equation 2.47. An important difference is that we have deliberately switched the role of the analysis and synthesis filters. In [20, 21] the $\mathbf{A}$ matrix contained the *analysis* filter coefficients, while $\mathbf{H}$ contained the *synthesis* filter coefficients. The reason for this twist is that we want the synthesis coefficients to be the free parameters in the optimization. This is discussed in Section 8.2.1.

The great advantage of Nayebi's formulation for perfect reconstruction is that all matrices and vectors are finite. That was not the case in in the general system description in the previous sections.

A time domain algorithm for designing analysis/synthesis FIR filter banks was derived in [20, 21] and can be explained as follows. Rewrite Equation 2.47 as N matrix equations:

$$
\mathbf{A}\mathbf{h}_i = \mathbf{b}_i, \qquad i = 0, \ldots, N-1, \tag{2.48}
$$

where $\mathbf{h}_i$ and $\mathbf{b}_i$ are the ith columns of $\mathbf{H}$ and $\mathbf{B}$, respectively. Note that for $L > N$ (i.e. $p > 1$) the matrix equation is over-constrained: The $\mathbf{A}$

matrix has dimensions $(2p-1)N \times pN$, thus the number of scalar equations is $(2p-1)N$, whereas the number of variables in $\mathbf{h}_i$ is pN.

Denote by $\mathcal{R}_A$ the subspace of $\mathbf{R}^{(2p-1)N}$ spanned by the columns of $\mathbf{A}$. It is often referred to as the range space of $\mathbf{A}$. From linear algebra, it is well known that Equation 2.48 is solvable if and only if $\mathbf{b}_i$ is in the range space of $\mathbf{A}$ [23]. To design an exact analysis/synthesis system we must therefore construct the matrix $\mathbf{A}$ such that

$$\mathbf{b}_i \in \mathcal{R}_A, \qquad i = 0, \dots, N-1. \qquad (2.49)$$

For a given $\mathbf{A}$ matrix one can tell how well it performs in a filter bank setting by computing the projection of each $\mathbf{b}_i$ onto $\mathcal{R}_A$ and finding the error vector:

$$\mathbf{e}_i = \mathbf{b}_i - \mathcal{P}_A(\mathbf{b}_i), \qquad i = 0, \dots, N-1. \qquad (2.50)$$

The projection operator $\mathcal{P}_A(\cdot)$ can be expressed in matrix form as [23]:

$$\mathcal{P}_A(\mathbf{z}) = \mathbf{A}(\mathbf{A}^T\mathbf{A})^{-1}\mathbf{A}^T\mathbf{z}, \qquad \mathbf{z} \in \mathbf{R}^{(2p-1)N}. \qquad (2.51)$$

Using matrix notation, all error vectors, now regarded as columns of the matrix $\mathbf{E}$, can be evaluated as

$$\mathbf{E} = \mathbf{B} - \mathbf{A}(\mathbf{A}^T\mathbf{A})^{-1}\mathbf{A}^T\mathbf{B}. \qquad (2.52)$$

The analysis filter bank that realizes the minimal error vector norms is given by

$$\mathbf{h}_i = (\mathbf{A}^T\mathbf{A})^{-1}\mathbf{A}^T\mathbf{b}_i, \qquad (2.53)$$

or, in matrix form as

$$\mathbf{H} = (\mathbf{A}^T\mathbf{A})^{-1}\mathbf{A}^T\mathbf{B}. \qquad (2.54)$$

As a measure of the reconstruction error ε_P, using a synthesis filter bank $\mathbf{G}$ and an analysis filter bank given by Equation 2.54, we use the sum of the squared norms of the individual projection error vectors:

$$\varepsilon_P = \sum_{i=0}^{N-1} \|\mathbf{e}_i\|^2. \qquad (2.55)$$

Using this error criterion in conjunction with an iterative gradient search algorithm, we ensure that the filter taps will be tuned for perfect reconstruction. Additional error functions may also be included for optimization purposes, and that is the topic of Section 8.2.

2.6 Summary

The main topic of this chapter has been signal decomposition with the aim of minimum representation with implementable inverse. We have shown the relation between series expansion and multirate filter banks.

The matrix description of the expansion offers a very flexible way of signal representation. Although we have limited the discussion to the fixed filter bank case, the method also lends itself to describing systems where the basis vectors vary as a function of the local image statistics. This could be used for adaptive systems if accompanied by a method for basis vector updating.

The PR condition was also derived from multirate equations for a parallel filter bank. This did not reveal any new theoretical results, but is a simple alternative derivation.

The main result from this chapter is the necessary and sufficient condition for PR in a minimum representation based on filter bank decomposition, namely the mutual orthogonality conditions between the unit sample responses of the analysis and synthesis filter banks, including orthogonality also after relative shifts of kN samples ($k \in \mathbf{Z}$), when N is the number of channels. This property is usually denoted *biorthogonality*.

In the unitary case the situation is fairly simple. Then the analysis system matrix is composed of orthonormal basis vectors. The reconstruction matrix is then simply the matrix Hermitian. Analysis basis vectors of a given length then imply synthesis vectors of the same length. For the nonunitary case there is no guarantee that the synthesis vectors will even be of finite length for FIR analysis vectors. It is, however, possible to force the analysis and synthesis vectors to have the same length through an iterative design method which was included in this chapter as a background for the following discussion, but also as a practical design tool for the most general filter banks.

In the next chapter we introduce other conditions for PR systems and structures for filter banks that lend themselves to practical implementation. Many of the presented systems are *structurally constrained* PR systems, i.e. systems guaranteeing perfect reconstruction irrespective of the choice of filter coefficients in either the analysis or synthesis filter bank.

Chapter 3

Filter Bank Structures

Multirate filter banks play a key role in subband coding of images. As discussed in Chapter 1, the analysis filter bank decomposes the signal into frequency bins and through decimation modulates each channel down to baseband representation. However, the main objective is decorrelation of the source signal. After decomposition each signal component should thus carry as much unique information as possible. If this is obtained, simple scalar quantization can be used for signal compression without too much loss of compression efficiency. The decorrelation between channels can be fully guaranteed by ideal frequency separation in the filter bank. However, as will be shown in Chapters 4 and 5, full decorrelation can be obtained by letting the number of channels go to infinity, or if the input signal is stationary, by intra-band decorrelation through channel whitening.

The role of the interpolating synthesis filters is to reconstruct the original signal, or make a best approximation of the input signal based on quantized subband signals. The filter bank operations include bandpass filtering of the correctly positioned image component of each band. As we shall learn in Chapter 8, the ability to suppress such artifacts as *blocking* and *ringing* rests with the synthesis filter bank.

In this chapter we shall only consider the perfect reconstruction requirement in the analysis-synthesis filter bank combination. We return to the filter bank optimization problem in Chapters 5 and 8.

Let us briefly review some of the history of subband coder filter banks. Most of the theory goes back to speech coding systems. The first subband coding attempts were made with filter banks using nonoverlapping passbands

49

[25], thus avoiding the aliasing problem at the cost of discarding certain frequency regions. This design philosophy is, of course, disastrous except when applying very steep filters, which on the other hand should be avoided for reasons of ringing artifacts and coping with nonstationarities in the signals. The first breakthrough for designing aliasing-free two-channel filter banks was made by Esteban and Galand [16]. The employed filters were of the FIR type. A generalization to IIR filtering was next introduced in [26, 27]. The first two-channel filter bank with *perfect reconstruction* was developed by Smith and Barnwell [27]. All these two-channel filter banks were used in a tree-structure to obtain N-channel systems. In 1983 Rothweiler [28] presented a parallel polyphase filter bank with approximate aliasing cancellation. The next year the aliasing matrix method for perfect reconstruction systems was introduced [24]. A very important step forward was the two-channel FIR perfect reconstruction structures based on lattice structures [29]. Whereas the PR condition was earlier obtained by optimization, the new structure guaranteed PR irrespective of the coefficient selection. PR IIR systems based on allpass filters were first derived for two channels [30] and later generalized to parallel systems [31, 32]. Stable implementation of the reconstruction filters depends on the use of *anticausal filters*.

More recent filter banks like the cosine-modulated filter banks, independently derived by three researchers [33, 34, 35], give very low computational complexity, while pure parallel filter banks, that have to be optimized for PR, renders more flexibility than any other filter banks [21, 36] at the cost of higher complexity. Wavelet filter banks (Wavelet transforms) belong to the more recent developments [37], but it is unclear whether their *discretized* forms really represent novel filter systems. Wavelet descriptions have offered new design methods, some more insight in certain filtering aspects, a link to analog processing, and new names to known techniques.

Contents

In this chapter we wish to review some of the filter bank systems that we presently feel have some importance. We do not intend to be complete, nor present all important aspects of multirate filter banks. But we do believe that further development of image and video coding systems will partly depend on the merits of these filter banks.

It is quite clear that the hardest competitor to all filter banks is the *discrete cosine transform* (DCT) [38]. All present image and video coding standards are based on this transform. We will show in Chapter 5 that filter banks have a potential of higher coding gain. Also, as will be presented in Chapter 8, the visual artifacts can be reduced by using properly selected and optimized filters. Ease of implementation is a further issue. Many researchers have believed that filter banks are much more complex for hardware implementation than the DCT. The fact is that filter banks which have a simpler implementation than the DCT exist. The most promising filter banks in terms of coding performance may be slightly more complex. As VLSI technology develops further, the complexity issues should be of little concern.

The organization of the chapter is as follows: The first two sections cover z-domain PR conditions. The following section describes two-band systems and their merits as building blocks of tree-structured systems. The last section is devoted to different parallel filter banks. We will mainly consider PR systems, but include some important exceptions. Both FIR and IIR filter systems are reviewed.

3.1 z-Domain Notation

In Chapter 2 a spatial domain description of filter banks was given using elements from linear algebra as mathematical tools. In the present chapter transform domain techniques are provided.

In the frequency domain a filter bank is supposed to partition the signal spectrum in several frequency bins as depicted in Figure 3.1. The filter bank

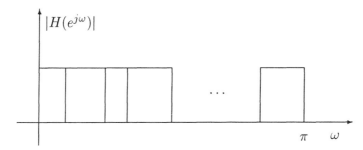

Figure 3.1: *Frequency bins of a filter bank illustrating the partitioning of the input signal spectrum.*

in this figure is *nonuniform*, that is, the filters have different bandwidths.

The filter bank is also *ideal* in the sense that the filters are nonoverlapping and contiguous. These properties prevent repeated representation of frequency components, as would be the case for overlapping filters, and also avoid loss of frequency components because the whole frequency range is covered.

We are going to assume that there are N channels in the filter bank. The transfer function of channel no. n is denoted by $H_n(z)$. If the transfer functions of the different channels are arranged in a column vector as

$$\mathbf{h}(z) = [H_0(z), H_1(z), \ldots, H_{N-1}(z)]^T, \tag{3.1}$$

and the output signals from the analysis filter bank are arranged in a similar column vector (in z-domain notation)

$$\mathbf{y}(z) = [Y_0(z), Y_1(z), \ldots, Y_{N-1}(z)]^T, \tag{3.2}$$

then the complete system can be compactly expressed in vector notation as

$$\mathbf{y}(z) = \mathbf{h}(z)X(z), \tag{3.3}$$

where $X(z)$ is the z-domain version of the scalar input signal.

The above equation does not include the necessary downsampling process, which is crucial in subband coding to make a minimum representation prior to quantization. Appendix A gives an overview of the mathematical mechanisms experienced in the downsampling as well as the upsampling process. The implications thereof in filter bank design when perfect reconstruction in the cascade of an analysis and a synthesis filter bank is a key issue, is the main topic of the following sections. In fact, the next two sections introduce two techniques for imposing PR on a subband filter bank system.

3.2 z-Domain PR Filter Bank Conditions

In this section we will present a PR requirement in the z-domain for a *uniform* filter bank where the output signals are critically sampled. This will shed some new light on the structure of the synthesis filter bank when the analysis filter bank is given.

The filter bank and the pertinent z-domain notation for signals and operators in the analysis filter bank is shown in Figure 3.2. Critical sampling

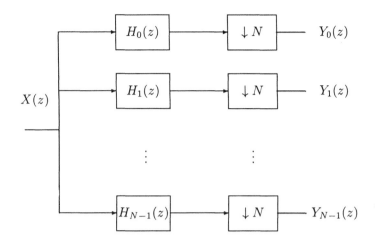

Figure 3.2: *Uniform decimating filter bank.*

in a uniform filter bank means that the downsampling factor is equal to the number of channels N. In general it is necessary to combine modulation and downsampling to obtain critically sampled lowpass representations. When using filters with ideal responses arranged as in Figure 3.3, the modulation to baseband is performed by downsampling.

With the above assumptions and the use of the multirate relations from Appendix A, the input-output relation of channel no. n, of this filter bank can be expressed as

$$Y_n(z) = \frac{1}{N} \sum_{m=0}^{N-1} H_n(z^{1/N} e^{-j\frac{2\pi}{N}m}) X(z^{1/N} e^{-j\frac{2\pi}{N}m}), \qquad (3.4)$$

where $H_n(z)X(z)$ is the signal in channel no. n prior to downsampling. The left-hand side of this equation can also be expressed in terms of z^N, for which case the signal relations are expressed as

$$Y_n(z^N) = \frac{1}{N} \sum_{m=0}^{N-1} H_n(ze^{-j\frac{2\pi}{N}m}) X(ze^{-j\frac{2\pi}{N}m}). \qquad (3.5)$$

We put these relations into matrix form by defining the elements of the column vector $\mathbf{x}(z)$ and the *alias component matrix* $\mathbf{H}(z)$ as

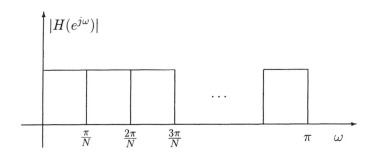

Figure 3.3: *Frequency partitioning in an ideal filter bank suited for integer band downsampling for obtaining baseband representation.*

$$x_m(z) \quad = \quad X(ze^{-j\frac{2\pi}{N}m}) \tag{3.6}$$

$$H_{nm}(z) \quad = \quad H_n(ze^{-j\frac{2\pi}{N}m}). \tag{3.7}$$

Also collecting the output signals in the column vector $\mathbf{y}(z^N)$, Equation 3.5 can be reformulated as

$$\mathbf{y}(z^N) = \frac{1}{N}\mathbf{H}(z)\mathbf{x}(z). \tag{3.8}$$

If $\mathbf{H}(z)$ is nonsingular, this equation can be inverted for recovery of $\mathbf{x}(z)$ from $\mathbf{y}(z^N)$ as

$$\mathbf{x}(z) = N\mathbf{H}^{-1}(z)\mathbf{y}(z^N) = \mathbf{G}(z)\mathbf{y}(z^N). \tag{3.9}$$

The inverse operator $\mathbf{G}(z) = N\mathbf{H}^{-1}(z)$ represents the synthesis filter.

Synthesis Filter Bank

Actually, the original signal is given by the first element of the vector $\mathbf{x}(z)$, that is, the signal reconstruction is given by

$$X(z) = x_0(z) = \sum_{i=0}^{N-1} G_{0i}(z)Y_i(z^N). \tag{3.10}$$

To clarify the situation further, the matrix element $G_{0i}(z)$ can be expressed as

$$G_{0i}(z) = \frac{N\hat{G}_i(z)}{\det \mathbf{H}(z)},$$ (3.11)

where $\hat{G}_i(z)$ is the *cofactor* of the matrix element $H_{0i}(z)$ [23]. With this definition Equation 3.10 can be written as

$$X(z) = x_0(z) = \frac{N}{\det \mathbf{H}(z)} \sum_{i=0}^{N-1} \hat{G}_i(z)Y_i(z^N).$$ (3.12)

Based on Equation 3.12 we can implement the synthesis filter bank as shown in Figure 3.4.

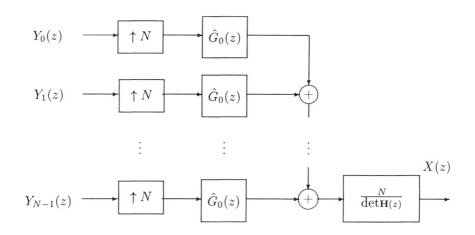

Figure 3.4: *Reconstruction filter bank.*

The synthesis filter bank structure in Figure 3.4 has a couple of important implications. With the analysis filter bank given, the perfect reconstruction property is guaranteed through a two-stage structure. The common filter block containing the inverse determinant of the analysis filter matrix compensates for *linear* distortion in the overall transfer function. Therefore, aliasing cancelation (see Appendix A) is obtained by selecting the channel filters as the cofactors of the analysis system matrix as defined in Equation 3.11.

In general, the filter order of the reconstruction filters appears to become considerably higher than the order of the analysis filters, because the cofactors are products of several matrix elements. This theory also indicates that FIR analysis filters do not guarantee FIR synthesis filters. The system determinant will in such cases represent an all-pole IIR filter with high order. A design objective might therefore be to eliminate the determinant (that is, make it equal to a constant delay) through proper choice of analysis filters.

We have earlier seen that by selecting a *unitary* system transform, the synthesis filter bank utilizes the same filters as the analysis filters. The determinant is then, in fact, given by a pure delay and the cofactors, which are the synthesis filters, have the same complexity as the analysis filters. We can conclude that the selection of analysis filters is of utmost importance for obtaining low complexity synthesis filters, and vice versa.

3.3 Polyphase Filter Bank Structures

In this section we will impose some structure on the analysis and synthesis filter banks in form of the *polyphase description* introduced by Bellanger et al. [39]. This structure is used for realizing most parallel systems including two-channel banks.

Polyphase Decomposition of FIR and IIR Filters

The polyphase decomposition can be performed on any transfer function $H(z)$ according to

$$H(z) = \sum_{k=0}^{N-1} P_k(z^N) z^{-k}, \tag{3.13}$$

where $P_k(z)$ is called the kth *polyphase component* of the filter. An implementation of this equation is shown in Figure 3.5.

For an FIR filter the decomposition is straight forward. Assume that the unit sample response of a filter is given by $\{h(l),\ l = 0, 1, \cdots, L - 1\}$. Then the polyphase unit sample responses are given by $\{p_k(m) = h(k + mN),\ k = 0, 1, \cdots, N - 1,\ m = 1, 2, \cdots, \lfloor L/N \rfloor - 1\}$. The polyphase filter coefficients are thus obtained by uniform subsampling of the original filter coefficients at N different phases for the N subfilters [11].

For the case of IIR filters the decomposition is somewhat more involved. To show how the decomposition can be performed, we write the denominator

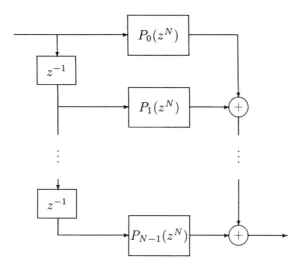

Figure 3.5: *Polyphase implementation of an FIR filter.*

of the transfer function in factored form,

$$H(z) = \frac{R(z)}{D(z)} = \frac{R(z)}{\prod_{j=1}^{K}(1 - \alpha_j z^{-1})}, \tag{3.14}$$

where $\{\alpha_j,\ j = 1, 2, \cdots, K\}$ are the filter poles. Let us modify one of the denominator factors as

$$\frac{1}{1 - \alpha_j z^{-1}} = \frac{1}{1 - \alpha_j z^{-1}} \frac{1 - \alpha_j^N z^{-N}}{1 - \alpha_j^N z^{-N}} = \frac{\sum_{i=0}^{N-1} \alpha_j^i z^{-i}}{1 - \alpha_j^N z^{-N}}. \tag{3.15}$$

In the two steps we have first multiplied the one-pole transfer function by a factor equal to 1. In the second step we have combined the numerator factor and one of the denominator factors, realizing that the combination can be expressed as the sum of a geometrical series with N terms.

By replacing all the denominator factors in the IIR transfer function in Equation 3.14 by the given equality, we obtain

$$H(z) = \frac{R(z) \prod_{j=1}^{K} \sum_{i=0}^{N-1} \alpha_j^i z^{-i}}{\prod_{j=1}^{K}(1 - \alpha_j^N z^{-N})} = \frac{S(z)}{D(z^N)}, \tag{3.16}$$

where the denominator is only a function of z^N. The numerator $S(z)$ represents the original zeros and the zeros generated by the poles.

It is straight forward to split this transfer function into polyphase components as

$$H(z) = \frac{1}{D(z^N)} \sum_{k=0}^{N-1} P_k(z^N) z^{-k}. \qquad (3.17)$$

Here, of course, the numerator has been split in the same way as was done for the FIR filter above. Note that all subfilters have the same denominator filter. A polyphase realization structure is shown in Figure 3.6.

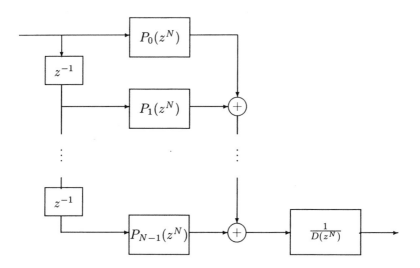

Figure 3.6: *Polyphase implementation of an IIR filter.*

An example might illustrate the mechanism in the transformation of an elliptic to a filter with the structure in Equation 3.16. The original filter zeros and poles are shown in Figure 3.7. The derived zeros and poles are shown in Figure 3.8.

In the IIR case, a notable alternative to the derivation of a filter with different z-dependency in the numerator and denominator from a standard filter is to design filters having the desired structure $S(z)/D(z^N)$ with frequency domain requirements [40, 41, 42]. This can lead to significantly better filters than obtained by the given method. A reason for this is that

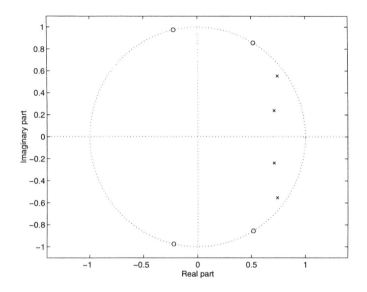

Figure 3.7: *Zeros and poles of an elliptic IIR filter.*

all the stopband zeros can be put on the unit circle rather than inside the unit circle as in the case of zeros generated from poles. Also, in the presented method, the pole-generated zeros appear periodically, a property which does not guarantee optimality.

Polyphase Filter Banks

Assume now that we have a set of N transfer functions

$$H_n(z), \ n = 0, 1, \cdots, N - 1 \tag{3.18}$$

representing filter bank channels. Split each filter in its polyphase components,

$$H_n(z) = \sum_{k=0}^{N-1} P_{nk}(z^N) z^{-k}. \tag{3.19}$$

Let us collect the different transfer functions in a column vector

$$\mathbf{h}(z) = [H_0(z), H_1(z), \cdots, H_{N-1}(z)]^T, \tag{3.20}$$

and the polyphase components in the polyphase matrix $\mathbf{P}(z^N)$ with elements $P_{nk}(z^N)$. The filter bank transfer functions can then be simultaneously

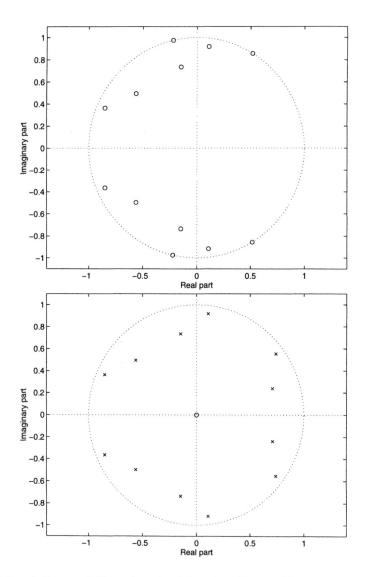

Figure 3.8: *a) Zeros of Equation 3.16 for the elliptic filter in Figure 3.7 for N = 3. b) Poles of the same filter.*

expressed as

$$\mathbf{h}(z) = \mathbf{P}(z^N) \begin{bmatrix} 1 \\ z^{-1} \\ \vdots \\ z^{-(N-1)} \end{bmatrix} = \mathbf{P}(z^N)\mathbf{d}_N(z), \qquad (3.21)$$

where $\mathbf{d}_N(z)$ is a vector of delay elements. This filter bank can be used as an analysis system where the N outputs $\mathbf{y}(z) = [Y_0(z), Y_1(z), \ldots, Y_{N-1}(z)]^T$ are calculated from $\mathbf{Y}(z) = \mathbf{h}(z)X(z)$. A polyphase realization is shown in Figure 3.9.

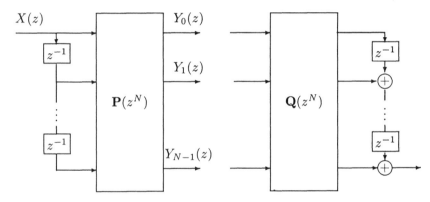

Figure 3.9: *Structure of an analysis filter bank (left) and a synthesis filter bank (right) realized in polyphase form. The polyphase matrices are called $\mathbf{P}$ and $\mathbf{Q}$, respectively. The two structures are related through filter transposition [11].*

Also shown in Figure 3.9 is a synthesis system where the polyphase matrix is denoted by $\mathbf{Q}(z^N)$. The synthesis system combines N channels $\mathbf{y}(z)$ through $X(z) = \mathbf{g}^T(z)\mathbf{y}(z)$. The complete synthesis filter bank can be described mathematically as

$$\mathbf{g}(z) = \mathbf{Q}^T(z^N)\tilde{\mathbf{d}}_N(z), \qquad (3.22)$$

where

$$\tilde{\mathbf{d}}_N^T(z) = [z^{-(N-1)}, z^{-(N-2)}, \ldots z^{-1}, 1]. \qquad (3.23)$$

Notice that $\tilde{\mathbf{d}}_N(z)$ is a reversed (upside down) version of $\mathbf{d}_N(z)$.

Critically Sampled Polyphase Systems

The polyphase decomposition described above results in a filter structure
that can be used to realize any filter. However, it lends itself particularly
well to multirate systems because of the z^N dependency of the polyphase
matrix.

In the subband coding context we must use critical downsampling by N
in the analysis channels, and upsampling by N in the synthesis channels.
With the aid of the noble identities (see Appendix A), we can move the
downsamplers and upsamplers across the polyphase matrices resulting in a
change of argument in the polyphase matrices from z^N to z. Illustrations of
the analysis and synthesis filter banks are given in Figure 3.10 after making
these changes.

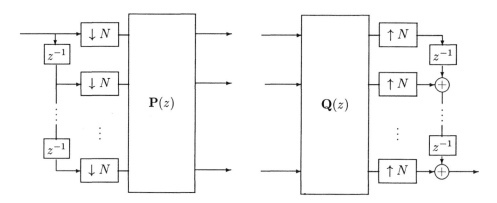

Figure 3.10: *Structure of a critically sampled analysis filter bank (left) using
polyphase form, and a similar interpolation synthesis filter bank (right).*

Now, using the two structures in a back-to-back configuration in an
analysis-synthesis setting, the condition for perfect reconstruction is

$$\mathbf{P}(z)\mathbf{Q}(z) = z^{-k_0}, \tag{3.24}$$

where k_0 is the system delay. A more general PR condition can be found in
[3].

The construction of the inverse filter (synthesis) bank depends on matrix inversion also in this case. Later we will, however, make the inversion

simple by adding some structure to the polyphase matrix . The most useful known building blocks for tractable implementations of filter banks are, as we shall see later, all-pass filters for IIR systems and lattice structures for FIR systems, possibly combined with Fourier type transforms.

It is important to make some comparisons between the polyphase and the alias matrix conditions for PR. In the alias matrix formulation we find direct conditions on the channel filters. In the polyphase matrix formulation the PR condition is imposed on the polyphase components. As we are generally dealing with matrix inversion also in the polyphase form, it is easily recognized that the inverse system may end up being of higher order than the analysis system. If the analysis filters are all of the FIR type, the synthesis filters may very well be IIR filters. This may also imply filter instability. We illustrate these aspect of inverse systems in several examples in this chapter.

Whenever we talk about filter or filter bank transfer functions, we shall always mean the transfer function *excluding* the downsamplers and upsamplers in multirate systems.

3.4 Two-Channel Filter Banks

Two-channel systems are of special interest as they can be most easily analyzed and optimized, and from an application point of view, they can form tree-structured systems as discussed in Section 3.5. They are also the basic building block in what is often called *wavelet transforms.*

The outline of this section is as follows: We investigate two-channel systems in light of the alias component matrix formulation and the polyphase matrix formulation leading to a variety of solutions. One filter system combination is that of *quadrature mirror filters* (*QMF*). An FIR QMF system guarantees aliasing-free reconstruction, but does not give perfect reconstruction. This filter bank type was the first successful filter bank presented in the literature [16]. The IIR counterpart based on allpass filters, however, does give perfect reconstruction if the combination of causal and anticausal filters is applied in the analysis and synthesis parts, respectively. The last, and perhaps most applicable two-channel filter type is based on lattice structures. These filters are in polyphase form, and therefore the polyphase matrix theory comes in handy.

3.4.1   Alias Component Matrix Description

In this section the alias component matrix formulation in Section 3.2 is specialized to two channels. For clarity we denote the analysis filter transfer functions in the two-channel case by $H_{LP}(z)$ and $H_{HP}(z)$ for the lowpass and highpass channels, respectively[1]. The alias component matrix for a two-band system is found as

$$\mathbf{H}(z) = \left[\begin{array}{cc} H_{LP}(z) & H_{HP}(z) \\ H_{LP}(-z) & H_{HP}(-z) \end{array} \right].$$ (3.25)

According to Figure 3.4 the synthesis filter bank for a two-channel system in general has a structure composed of channel filters and a fullband equalizer. From Equation 3.11 the synthesis channel filters are found as

$$G_{LP}(z) = \frac{2}{\det \mathbf{H}(z)} H_{HP}(-z)$$ (3.26)

and

$$G_{HP}(z) = -\frac{2}{\det \mathbf{H}(z)} H_{LP}(-z).$$ (3.27)

The amplitude equalization term is here given by

$$\frac{2}{\det \mathbf{H}(z)} = \frac{2}{H_{LP}(z)H_{HP}(-z) - H_{LP}(-z)H_{HP}(z)} = \frac{2}{F(z) - F(-z)}.$$
(3.28)

A filter configuration for a two-channel PR system is given in Figure 3.11.

This model is general enough to cover all possible special cases. If the analysis filter bank is of the FIR type, then the synthesis filter bank will in general be of the IIR type. However, one may force the synthesis filter bank to be FIR by making the determinant equal to, or in some cases approximately equal to a constant plus a pure delay, that is

$$F(z) - F(-z) = 2z^{-k_0},$$ (3.29)

where k_0 is the necessary system delay, measured in number of samples. In that case, we see from the above equations that the synthesis filter bank will contain the same filters as the analysis bank except for an exchange of

---

[1]In earlier sections these filters were denoted by $H_0(z)$ and $H_1(z)$.

a) b)

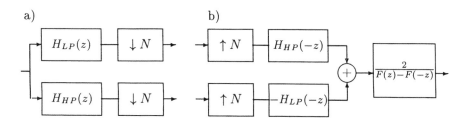

Figure 3.11: *a) Two band parallel analysis filter bank. b) Corresponding synthesis filter bank.*

responses and a replacement of z by $-z$, plus a sign change in the synthesis highpass filter (see Equation 3.27). The negation of the argument is a modulation corresponding to a lowpass-to-highpass transformation. In the time domain the relevant relation is $g_{LP}(k) = (-1)^k h_{HP}(k)$. In Figure 3.12 the frequency response relations of the filters are illustrated for an idealized filter example.

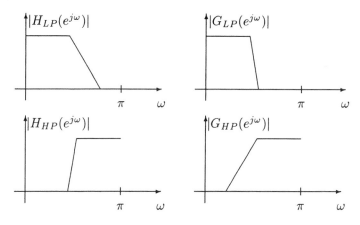

Figure 3.12: *Frequency responses of the analysis filters (left) and the synthesis filters (right) of a two-channel system.*

Let us analyze the consequences of the pure delay condition in Equation 3.29 further. Assume that $F(z)$ is given by

$$F(z) = \sum_{l=0}^{L-1} f(l) z^{-l}. \tag{3.30}$$

Then the determinant becomes

$$\det \mathbf{H}(z) = F(z) - F(-z) = \sum_{l=0}^{L-1} \left(1 - (-1)^l\right) f(l) z^{-l}. \qquad (3.31)$$

To force the determinant to become equal to a pure delay times 2, every second coefficient of $F(z)$ can be selected freely, one remaining coefficient in Equation 3.31 needs to be nonzero, while the others have to be zero. This can be obtained either as a structural constraint as in the lattice filter banks to be described shortly, or as a condition during filter bank optimization.

3.4.2   Polyphase Matrix Description

The polyphase matrix description of the two-channel system is

$$\mathbf{h}(z) = \left[\begin{array}{c} H_{LP}(z) \\ H_{HP}(z) \end{array} \right] = \left[\begin{array}{cc} P_{00}(z^2) & P_{01}(z^2) \\ P_{10}(z^2) & P_{11}(z^2) \end{array} \right] \left[\begin{array}{c} 1 \\ z^{-1} \end{array} \right] = \mathbf{P}(z^2) \mathbf{d}_2(z).$$

$$(3.32)$$

The filter structure for this equation including critical sampling is depicted in Figure 3.13. When $\mathbf{P}(z)$ is the (decimated) polyphase analysis matrix,

a)   b)

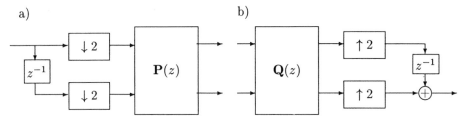

Figure 3.13: *General two-channel analysis polyphase structure (left) and corresponding synthesis structure (right).*

perfect reconstruction requires that

$$\mathbf{Q}(z) = \mathbf{P}(z)^{-1} = \frac{1}{\det \mathbf{P}(z)} \left[\begin{array}{cc} P_{11}(z) & -P_{01}(z) \\ -P_{10}(z) & P_{00}(z) \end{array} \right], \qquad (3.33)$$

where the determinant is given by

$$\det \mathbf{P}(z) = P_{00}(z) P_{11}(z) - P_{01}(z) P_{10}(z). \qquad (3.34)$$

We may also express the determinant of the alias component matrix presented in Equation 3.28 in terms of the polyphase components:

$$\det \mathbf{H}(z) = -2z^{-1} \left[P_{00}(z^2)P_{11}(z^2) - P_{01}(z^2)P_{10}(z^2) \right]. \qquad (3.35)$$

From the two equations above we find the relation between the determinants of the alias component matrix and the polyphase matrix

$$\det \mathbf{H}(z) = -2z^{-1} \det \mathbf{P}(z^2). \qquad (3.36)$$

As we can clearly see, in the two-channel case the requirements for perfect reconstruction without implementing the determinants are similar, as expected.

Instead of discussing the general case any further, we shall study several special cases in greater detail.

3.4.3   Quadrature Mirror Filters (QMFs)

A lowpass-highpass pair of filters obeying the relation

$$H_{HP}(z) = H_{LP}(-z) \qquad (3.37)$$

is called *quadrature mirror filters* (QMF). To appreciate the term QMF we write the relation between the frequency magnitude responses of the two filters

$$|H_{HP}(e^{j\omega})| = |H_{LP}(e^{j(\pi+\omega)})|, \qquad (3.38)$$

which exhibit mirror symmetry of the two responses around $\omega = \pi/2$ ($f = 0.25$). An example of such a magnitude relation is shown in Figure 3.14.

Using the inverse z-transform, we find the corresponding signal domain relation as

$$h_{HP}(l) = h_{LP}(l)(-1)^l. \qquad (3.39)$$

Thus the highpass filter is obtained from the lowpass filter through modulation of the unit sample response by the sequence $(-1)^l$.

Writing the z-domain expressions for the quadrature mirror filters in terms of their polyphase components, we get

$$H_{LP/HP}(z) = P_{00}(z^2) \pm z^{-1} P_{01}(z^2), \qquad (3.40)$$

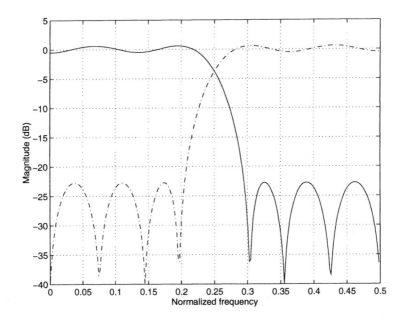

Figure 3.14: *Magnitude relation between FIR quadrature mirror filters designed using the Remez algorithm [43]. The filters have 16 coefficients.*

where the upper sign is valid for the lowpass filter and the lower sign for the highpass filter. Because only two polyphase components express both filters, we simplify the notation by introducing

$$P_0(z) = P_{00}(z),$$
$$P_1(z) = P_{01}(z).$$

The polyphase matrix is thus

$$\mathbf{P}(z) = \begin{bmatrix} P_0(z) & P_1(z) \\ P_0(z) & -P_1(z) \end{bmatrix} = \begin{bmatrix} 1 & 1 \\ 1 & -1 \end{bmatrix} \begin{bmatrix} P_0(z) & 0 \\ 0 & P_1(z) \end{bmatrix}. \qquad (3.41)$$

In this case the polyphase matrix has been decomposed into a cascade of the well known *Hadamard matrix*, which accounts for the special relationship between the lowpass and highpass filters, and a diagonal matrix including the polyphase components. The polyphase structure of the given QMF system is shown in Figure 3.15.

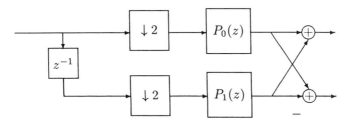

Figure 3.15: *Polyphase realization of the two-channel QMF bank.*

General Quadrature Mirror Filter Bank Inverse

The inverse of the quadrature mirror two-channel filter bank can easily be derived either by matrix inversions or simply by inspection. The inverse of the polyphase matrix is given by

$$\mathbf{Q}(z) = \mathbf{P}^{-1}(z) = \frac{1}{2} \begin{bmatrix} 1/P_0(z) & 0 \\ 0 & 1/P_1(z) \end{bmatrix} \begin{bmatrix} 1 & 1 \\ 1 & -1 \end{bmatrix}. \tag{3.42}$$

Figure 3.16 shows the corresponding synthesis filter bank with the PR condition satisfied.

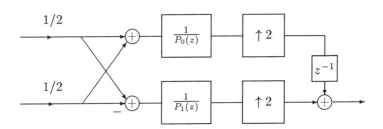

Figure 3.16: *Realization of the polyphase two-channel synthesis structure.*

The simplicity of verifying the PR property of the given analysis-synthesis system does not imply anything about its applicability. Let us first consider the case when the polyphase analysis filters are of the FIR type. As the

synthesis filters are inverse filters, they become allpole filters. The stability
of IIR filters requires that all poles be inside the unit circle. The implication on the analysis filters is that the polyphase filter components must
be minimum phase in the strict sense, that is, all zeros must be inside the
unit circle. To get further insight into this problem, let us illustrate with an
example.

Design Example

In this simple nontrivial example, the analysis polyphase components are
given by:

$$P_0(z) = 1 + \alpha z^{-1}, \quad \text{and} \quad P_1(z) = \beta + \gamma z^{-1}.$$

The explicit forms of the transfer functions for the analysis and synthesis
filters then become

$$H_{LP/HP}(z) = 1 \pm \beta z^{-1} + \alpha z^{-2} \pm \gamma z^{-3}, \tag{3.43}$$

and

$$G_{LP/HP}(z) = \frac{1}{2} \frac{1 \pm \beta z^{-1} + \alpha z^{-2} \pm \gamma z^{-3}}{(1 + \alpha z^{-2})(\beta + \gamma z^{-2})}. \tag{3.44}$$

We observe that both lowpass transfer functions have the same zeros, and
so do the highpass filters too. To obtain stable filters the poles have to be
inside the unit circle, implying that

$$\sqrt{\alpha} < 1 \tag{3.45}$$

and

$$\sqrt{\gamma/\beta} < 1. \tag{3.46}$$

On the other hand, to obtain good stopband attenuation, the zeros should be
on the unit circle. In this simple case, as there are three zeros, one should be
located at $z = -1$ and the two others should appear in a complex conjugate
pair at $\exp(\pm j\theta)$ where θ is the angular frequency of the zero corresponding
to the positive sign. $H(z)$ must accordingly be a symmetric polynomial
implying that $\gamma = 1$ and $\alpha = \beta$. The stability conditions then require that
$\alpha < 1$ and $1/\alpha < 1$. These two conditions are obviously contradictory.

□

There are two solutions for resolving the contradiction revealed in the example:

1. We can give up the linear phase requirement on $H(z)$ by moving the stopband zeros off the unit circle, thus lowering the stopband attenuation.

2. We can give up the stability requirement for the synthesis filters.

We now look at the second possibility.

We have here used the term *stability* rather loosely, assuming that the filter is causal. However, a filter with all poles outside the unit circle is stable if implemented in an anticausal fashion. In Appendix A anticausal filtering is discussed. For the present structure a possible procedure for obtaining useful filtering is the following:

- Design any prototype FIR lowpass analysis filter, where the zeros of the polyphase components must be off the unit circle.

- Split the prototype filter in its two polyphase components.

- Factorize the polyphase subfilters.

- Group all factors in each polyphase component with zeros outside the unit circle in $P_{a,i}(z)$, and those inside are grouped in $P_{c,i}(z)$. The indices represent: a: anticausal and c: causal.

- The anticausal filters are replaced by causal filters with the reciprocal poles: $P_{a,i}(z) \rightarrow P_{ar,i}(z) = P_{a,i}(z^{-1})$.

- The actual filtering in the cascade of $P_{c,i}(z)$ and $P_{ar,i}$ is performed as

  - $P_{ar,i}$ operates on reversed signals and the output is again reversed.

  - $P_{c,i}(z)$ operates in normal order.

Example Continued

When we accept poles outside the unit circle, we can satisfy the linear phase condition for $H_{LP/HP}(z)$ and for the numerator of $G_{LP/HP}(z)$. We then have the following two transfer functions:

$$H_{LP/HP}(z) = 1 \pm \alpha z^{-1} + \alpha z^{-2} \pm z^{-3} \qquad (3.47)$$

and

$$G_{LP/HP}(z) = \frac{1}{2} \frac{1 \pm \alpha z^{-1} + \alpha z^{-2} \pm z^{-3}}{(1 + \alpha z^{-2})(\alpha + z^{-2})}. \qquad (3.48)$$

The poles in $G_{LP/HP}(z)$ are now located at $\pm j\sqrt{\alpha}$ and $\pm j\sqrt{1/\alpha}$. α is related to the relative angular frequency of one of the complex conjugated pairs of zeros as $\alpha = 1 - 2\cos(\theta)$. In this particular case all filters are completely specified by the single parameter α. The pole-zero configuration of the synthesis filters are shown in Figure 3.17. The analysis filter zeros are equal to the synthesis filter zeros.

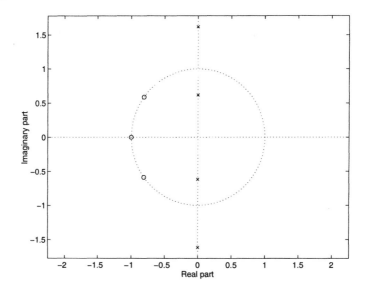

Figure 3.17: *Pole-zero plot for the synthesis filter in the example. The parameter $\theta = 0.4$.*

Typical frequency responses for the analysis and synthesis lowpass filters are shown in Figure 3.18.

□

The derived filter bank is definitely not *unitary*, as can be verified from the different magnitude responses of the analysis and synthesis lowpass filters

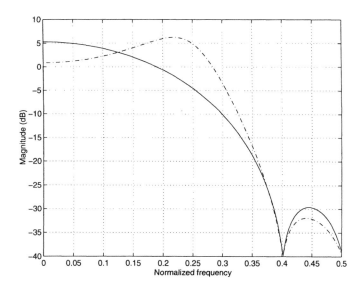

Figure 3.18: *Frequency responses of analysis (solid line) and synthesis (dashed-dotted line) lowpass filters in the noncausal filter configuration. The parameter $\theta = 0.4$.*

in Figure 3.18. Recall from Chapter 2 that for unitary systems the analysis and synthesis unit sample responses in a given channel are reversed versions of each other. This implies that the magnitudes of the frequency responses be equal.

The required signal reversal may be awkward in practical applications. The reversal has to be done either on the analysis or the synthesis side. In broadcasting applications using this kind of filter bank, it may also be advantageous to interchange the analysis and synthesis filters to make the receiver simple. In Chapter 8 and Appendix B we will also point out the necessity of having strong limitations on the lengths of the unit sample responses in the synthesis filters. As IIR filters, by definition, have infinite length responses, the use of the FIR filter in the synthesis filter bank definitely seems to be a good idea. Furthermore, the IIR-based frequency response can provide *whitening* within a channel and thus help maximizing the coding gain, as demonstrated in Chapter 5.

Allpass Polyphase Components

The use of IIR filters with both zeros and poles in the polyphase subfilters in the analysis filter bank is a straight forward extension of the system in the previous section. The synthesis polyphase subfilters would then have inverse transfer functions, i.e. zeros replaced by poles and vice versa. If we allow for both causal and anticausal filtering, this would be a very flexible design. Here we will consider one particular example of great practical importance. This system is motivated from the observation that the frequency responses of the polyphase filters have to be close to allpass filter responses to make PR systems.

Suppose we choose the polyphase subfilters exactly as allpass filters:

$$P_i(z) = A_i(z), \ \ 1 = 1, 2. \tag{3.49}$$

Using a property of allpass filters (see Appendix A), the necessary inverse polyphase filters then becomes

$$1/P_i(z) = 1/A_i(z) = A_i(z^{-1}). \tag{3.50}$$

Stability of the synthesis filters requires anticausal implementation. This simplifies the operations somewhat compared to combined causal and anticausal operations.

Another important issue is: can we make good filters based on allpass filters of the form

$$H_{LP/HP}(z) = \frac{1}{2} \left(A_0(z^2) \pm z^{-1} A_1(z^2) \right) ? \tag{3.51}$$

The answer is affirmative [44]. Let us consider the frequency response of the allpass-based filter bank. The frequency response of any allpass filter can be written

$$A(e^{j\omega}) = e^{j\phi(\omega)}, \tag{3.52}$$

where $\phi(\omega)$ is the phase of the filter and is a real function. Then the magnitude responses become

$$|H_{LP/HP}(e^{j\omega})|^2 = \frac{1}{2} \left[1 \pm \cos(\omega - \phi_1(2\omega) + \phi_0(2\omega)) \right], \tag{3.53}$$

where $\phi_0(\omega)$ and $\phi_1(\omega)$ are the phases of $A_0(e^{j\omega})$ and $A_1(e^{j\omega})$, respectively. It is evident that $|H_{LP/HP}(e^{j\omega})| \leq 1$, which implies that the response is structurally upper bounded by 1 [45], no matter how we select the filter coefficients. From Equation 3.53 we also easily derive that the lowpass and highpass filters are power complementary [45], meaning that

$$|H_{LP}(e^{j\omega})|^2 + |H_{HP}(e^{j\omega})|^2 = 1. \tag{3.54}$$

Thus, if $H_{LP}(z)$ is a lowpass filter, then $H_{HP}(z)$ will be the power complementary highpass filter.

Example

Let us consider a simple but nontrivial example. Choose the polyphase allpass filters as

$$A_0(z) = \frac{\alpha + z^{-1}}{1 + \alpha z^{-1}} \quad \text{and} \quad A_1(z) = 1. \tag{3.55}$$

Note that the only free parameter in this system is α. With these building blocks, the filter responses are given by

$$H_{LP/HP}(z) = \frac{1}{2}\left[\frac{\alpha + z^{-2}}{1 + \alpha z^{-2}} \pm z^{-1}\right] = \frac{1}{2}\frac{\alpha \pm z^{-1} + z^{-2} \pm \alpha z^{-3}}{1 + \alpha z^{-2}}. \tag{3.56}$$

The pole-zero plot for the lowpass filter for $\alpha = 0.382$ is shown in Figure 3.19 (top). The parameters in the figure are related to the filter parameters through: $\theta = \arccos\left[(1 - \alpha)/2\right]$ and $z_p = \pm j/\sqrt{\alpha}$. Typical frequency responses are shown in Figure 3.19 (bottom). The highpass filter responses can be found by mirroring the given responses at $f = 0.25$.

It is interesting to compare the allpass filter responses with those in Figure 3.18. The sum of the analysis and synthesis responses in the passband is approximately equal to 6 dB in both cases. 6 dB corresponds to the loss in the interpolator due to the insertion of zero samples between the original samples. Both filter systems rely on one parameter α only. The stopband behavior is close for both systems while there is a dramatic difference in the passband responses. The allpass based system has a very flat passband response, which cannot be matched by the FIR/IIR system. However, when we consider gain in subband coders in Chapter 5, we learn that the lift in the broken curve in Figure 3.18 is close to what optimality requires.

□

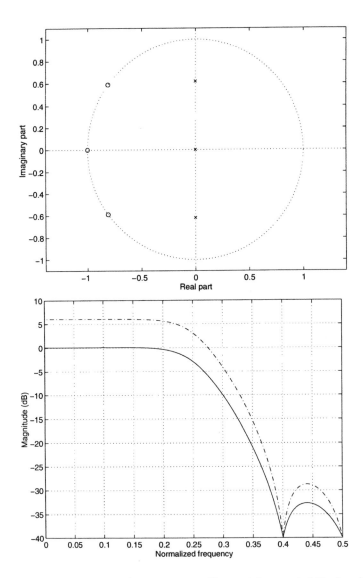

Figure 3.19: *Pole-zero plot of the lowpass filter with $\alpha = 0.382$ (top) and low-pass channel frequency responses of the analysis (broken line) and synthesis (solid line) filter of Equation 3.56 (bottom).*

Allpass filters can in general be implemented using approximately half the number of multiplications compared to a general equal order IIR filter. For the first order allpass filter in the above example, the multiplication rate is only one multiplication per input sample. This corresponds to half a multiplication per input sample in the decimating analysis filter bank. Several efficient first order allpass structures exist [46]. One example is the structure shown in Figure 3.20.

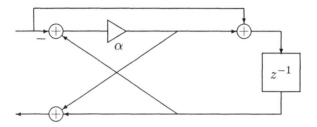

Figure 3.20: *One-multiplier first-order allpass structure.*

It is worth noticing that the sensitivity to coefficient quantization is low for the allpass based systems. This is particularly true for the passband responses due to the structural boundedness [47]. It often suffices to represent the coefficients by 2-3 nonzero digits for practical image decomposition filters. An excellent filter is obtained by using first order allpass filters in both branches with coefficients $0.625 = 2^{-1} + 2^{-3}$ and $0.125 = 2^{-3}$, respectively. It is easily recognized that the first multiplication can be performed by two shift operations and one addition, while the second multiplication requires one shift operation.

The two-channel allpass-filter based systems form unitary filters because they are PR systems and the synthesis unit sample responses are reversed versions of the analysis unit sample responses.

QMF Systems with FIR Analysis and Synthesis Filters

So far we have designed QMF systems where IIR filters are involved either in the analysis or synthesis filters, or even both. A much desired system would use FIR filters in both filter banks, possibly with the same filter lengths in the analysis and synthesis filters.

As mentioned earlier, the FIR quadrature mirrors filters of [16] were

the first practical subband filters to occur. Unfortunately they are not perfect reconstruction systems. All aliasing components are eliminated in the
synthesis filter bank, but the magnitude of the overall frequency response
through the combined analysis-synthesis system can never be exactly unity.
With increasing filter order the response can be made as close to perfect as
desired.

In the FIR QMF case we insist that the synthesis system have the same
structure as the analysis system with FIR filters of the same length as illustrated in Figure 3.21.

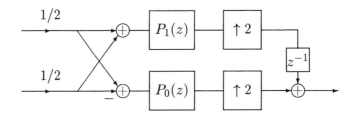

Figure 3.21: *Quadrature mirror approximate filter inverse.*

The design objective is

$$P_0(z)P_1(z) \approx z^{-k_0}, \tag{3.57}$$

where $P_0(z)$ and $P_1(z)$ are the polyphase components of the analysis filter.

Johnston [17] designed filters of this type intended for use in speech
coders. These have also enjoyed great popularity in the image coding community. We include a design example where the Johnston filter bank f8a is
demonstrated. The frequency responses of the lowpass and highpass analysis filters are shown in Figure 3.22 (top). Their coefficients are specified in
Appendix C. The synthesis filters are the same except for the scaling factor.
Figure 3.22 (bottom) shows the overall transfer function of the combined
analysis-synthesis filter banks.

In a practical setting, the nonperfect reconstruction property is not necessarily prohibitive. However, the PR property can be obtained without
sacrificing complexity, and it is therefore worth while to consider other alternative filter systems.

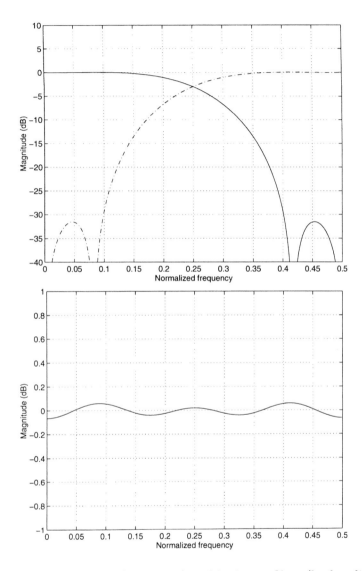

Figure 3.22: *Lowpass filter (solid line) and highpass filter (broken line) magnitude responses of Johnston type f8a two-channel filter bank (top), and magnitude response of a two-channel analysis/synthesis system based on the Johnston f8a filter (bottom).*

3.4.4   FIR PR Systems

Because the FIR QMF systems cannot yield perfect reconstruction, it is important to find other solutions. The first PR system to appear in the literature is, to our knowledge, the system introduced by Smith and Barnwell [48]. We therefore treat that method first. The second topic of this section is a general case where we require that the synthesis filters be FIR when the analysis filters are FIR. This is done through a simple example which give some extra insight and also leads to lattice structures. The lattice structures are probably the most important FIR PR structures from a practical point of view, and are presented last in this main section.

The Smith-Barnwell Method

In the method by Smith and Barnwell [48], the relation between the lowpass and the highpass analysis filters is selected as

$$H_{HP}(z) \ = H_{LP}(-z^{-1}). \tag{3.58}$$

This relation does not satisfy the QMF condition. The reconstruction filters can be found from Equations 3.26 and 3.27 as

$$G_{LP}(z) = \frac{2}{\det \mathbf{H}(z)} H_{LP}(z^{-1}), \tag{3.59}$$

and

$$G_{HP}(z) = -\frac{2}{\det \mathbf{H}(z)} H_{LP}(-z), \tag{3.60}$$

where $\det \mathbf{H}(z) = F(z) - F(-z)$ and $F(z) = H_{LP}(z)H_{LP}(z^{-1})$. We now try to simplify the above synthesis filters by selecting the analysis filters in such a way that $F(z) - F(-z) = 2z^{-(2k-1)}$.

If designing $F(z)$ as an FIR filter of length $L = 4k - 1$, $\quad k \in Z_+$, with unit sample response $f(n)$, $\ n = 0, 1, \ldots, 4k - 2$ satisfying:

$$f(n) = \begin{cases} 2 & \text{for } n = 2k - 1 \\ 0 & \text{for } n = 1, 3, \ldots, 4k - 3 \\ \text{arbitrary} & \text{for } n \text{ even}, \end{cases} \tag{3.61}$$

then, obviously, our goal is reached.

The requirements above are exactly those of a linear phase *half-band filter* with double zeros on the unit circle. $F(z)$ can then be factored in the two components with inverse relationships between the zeros. If $H_{LP}(z)$ has a zero at $z_n = \rho_n \exp(j\theta_n)$, then $H_{LP}(z^{-1})$ will have a zero at $z_n^{-1} = \rho_n^{-1} \exp(-j\theta_n)$. One possibility is therefore to split $F(z)$ in a minimum phase filter $H_{LP}(z)$ and a maximum phase filter $H_{LP}(z^{-1})$. $H_{LP}(z)$ would then contain all zeros inside the unit circle and one zero from each zero pair on the unit circle.

Design Method

The above observations give rise to a simple design technique for filters according to Smith and Barnwell satisfying the PR condition. The trick is to design a linear phase filter using the Remez algorithm [43], and thereafter manipulate the result using the technique suggested in [49].

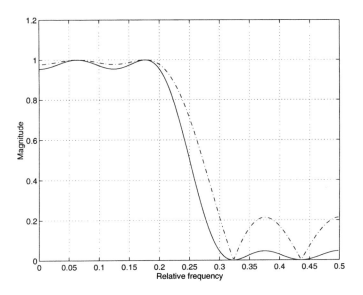

Figure 3.23: *Magnitude response (solid line) of the model half-band filter $F(z)$ to be used in the Barnwell Smith PR system. Magnitude response of the derived lowpass filter (broken line).*

In the figure both curves are normalized to make the maximum passband gain unity. The magnitude responses with the given normalization are related through:

$$|H(e^{j\omega})| = \sqrt{|F(e^{j\omega})|}.\tag{3.62}$$

Assume that a perfect equiripple FIR half-band filter has been designed. This filter has complete rotation-by-180-degrees symmetry in its *zero-phase* frequency response around the point $(f = 0.25,\quad H(e^{j\omega}) = 0.5)$. This also indicates that the maximum deviations from 1 in the passband and from 0 in the stopband are the same on a linear scale. Now, find the maximum of the frequency response, say $1+\delta$. By adding δ to the center tap (symmetry tap) of the filter, we move the frequency response upwards so that the minima are exactly at zero in the stopband. This means that two and two of the stopband zeros merge. We can then proceed to decompose $F(z)$ into the two filters as indicated above.

An example of the above procedure is given in Figure 3.23, where a filter of length 15 representing $F(z)$ and the resulting $H_{LP}(z)$ are shown.

To obtain unity gain through the cascade of the analysis and the synthesis filter banks the filters have to be rescaled.

The zeros of the two component filters are given in Figures 3.24 and 3.25.

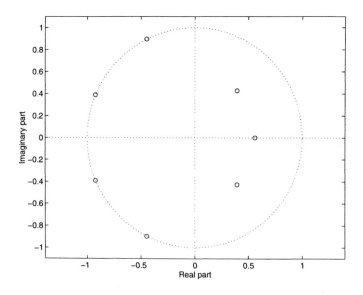

Figure 3.24: *Zeros of causal filter.*

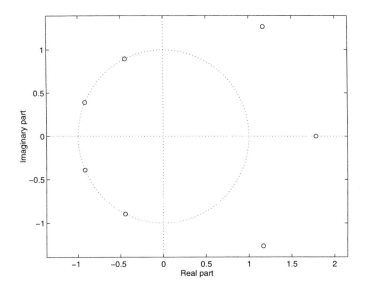

Figure 3.25: *Zeros of anticausal filter.*

FIR PR Systems

In this section we do not assume any specific relation between the analysis filters, but require perfect reconstruction using FIR filters only. We base the derivation on the polyphase formalism in Equation 3.32. An example given below introduces this topic. From the example we draw conclusions that generalize the two-channel FIR theory.

Filter Example

In this example the analysis system consists of length 4 FIR filters in the lowpass and highpass branches, respectively. We require that the synthesis filters be FIR of same length, as well. Without loss of generality we select monic polynomials for the transfer functions of the two filters:

$$H_{LP}(z) = 1 + h_1 z^{-1} + h_2 z^{-2} + h_3 z^{-3}, \tag{3.63}$$

and

$$H_{HP}(z) = 1 + k_1 z^{-1} + k_2 z^{-2} + k_3 z^{-3}. \tag{3.64}$$

The polyphase matrix then becomes

$$\mathbf{P}(z) = \begin{bmatrix} 1 + h_2 z^{-1} & h_1 + h_3 z^{-1} \\ 1 + k_2 z^{-1} & k_1 + k_3 z^{-1} \end{bmatrix}. \tag{3.65}$$

To make the inverse filter bank FIR, the polyphase matrix determinant must constitute a pure delay. For this case the determinant is equal to

$$\det \mathbf{P}(z) = (k_1 - h_1) + (k_3 - h_3 + h_2 k_1 - h_1 k_2) z^{-1} + (h_2 k_3 - h_3 k_2) z^{-2}. \tag{3.66}$$

We obviously have three choices for making the determinant equal to a delay.

Case 1

Requirements:

1. $k_1 - h_1 = 0$

2. $h_2 k_3 - h_3 k_2 = 0$.

If selecting the parameters of the lowpass filter and a new parameter d as free parameters, the highpass filter is then given by

$$H_{HP}(z) = 1 + h_1 z^{-1} + d(h_2 + h_3 z^{-1}) z^{-2}. \tag{3.67}$$

From this result we can restate the polyphase matrix as

$$\mathbf{P}(z) = \begin{bmatrix} 1 + h_2 z^{-1} & h_1 + h_3 z^{-1} \\ 1 + dh_2 z^{-1} & h_1 + dh_3 z^{-1} \end{bmatrix}. \tag{3.68}$$

We observe the strong coupling between the lowpass and highpass filters. Due to this coupling, we can realize the analysis filters as a lattice structure as shown in Figure 3.26. The synthesis filter bank is based on the polyphase matrix inverse, which can be expressed solely by the lowpass analysis filter parameters and d. From Equation 3.33 we find,

$$\mathbf{Q}(z) = \mathbf{P}^{-1}(z) = \frac{1}{(d-1)(h_3 - h_1 h_2)} \begin{bmatrix} h_1 + dh_3 z^{-1} & -h_1 - h_3 z^{-1} \\ -1 - dh_2 z^{-1} & 1 + h_2 z^{-1} \end{bmatrix}. \tag{3.69}$$

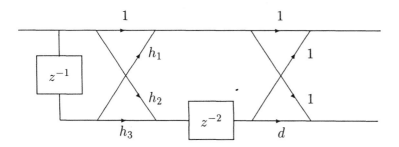

Figure 3.26: *Lattice analysis structure of example filter.*

The synthesis filter transfer functions are given by

$$\mathbf{g}^T(z) = \tilde{\mathbf{d}}_2(z)\mathbf{Q}(z^2),\tag{3.70}$$

which spells out explicitly for the example filter as

$$G_{LP}(z) = (-1 + h_1 z^{-1} - d(h_2 z^{-2} - h_3 z^{-3}))/((d-1)(h_3 - h_1 h_2)),\tag{3.71}$$

and

$$G_{HP}(z) = (1 - h_1 z^{-1} + h_2 z^{-2} - h_3 z^{-3})/((d-1)(h_3 - h_1 h_2)).\tag{3.72}$$

A possible synthesis filter realization structure is given in Figure 3.27.

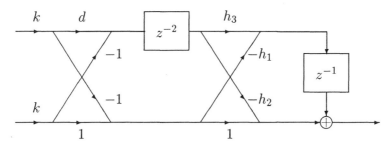

Figure 3.27: *Lattice synthesis structure of example filter.*
$k = 1/((d-1)(h_3 - h_1 h_2))$.

To illustrate the potential of a simple filter like the one above, we show a filter system where the analysis lowpass filter zeros are located at $z_{l_1} = -1$ and two zeros at $z_{l_{2,3}} = 0.2$. If the highpass filter parameter $d = 5$, we obtain highpass filter zeros at $z_{h_1} = 1$, $z_{h_2} = -1.7165$, and $z_{h_3} = 0.1165$.

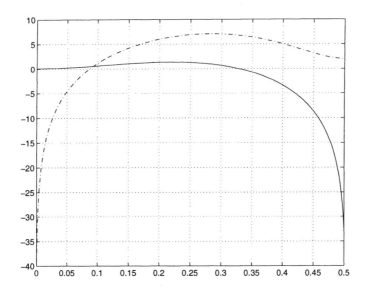

Figure 3.28: *Magnitude responses of lowpass (solid line) and highpass (dashed line) analysis filters of example.*

The magnitude responses of the analysis filters are shown in Figure 3.28. The synthesis filters are found from Equations 3.71 and 3.72.

It is possible to constrain the lowpass filter to be linear phase without doing the same for the highpass filter. The requirements are: $h_3 = 1$ and $h_2 = h_1$. Then the highpass filter becomes:

$$H_{HP} = 1 + h_1 z^{-1} + d(h_1 + z^{-1})z^{-2}. \tag{3.73}$$

If we also require linear phase for the highpass filter, there are two choices: $d = \pm 1$. The use of the positive sign leads to a degenerate case for which both filters are equal. With the minus sign we get meaningful filters:

$$H_{LP}(z) = 1 + h_1 z^{-1} + h_1 z^{-2} + z^{-3}, \tag{3.74}$$

and

$$H_{HP}(z) = 1 + h_1 z^{-1} - h_1 z^{-2} - z^{-3}. \tag{3.75}$$

The synthesis filter can be found from Equation 3.72.

Case 2

Requirements:

1. $h_2 k_3 - h_3 k_2 = 0$

2. $k_3 - h_3 + h_2 k_1 - h_1 k_2 = 0$

Eliminating k_2 and k_3 from these equations we can express the highpass analysis filter as:

$$H_{HP}(z) = 1 + k_1 z^{-1} + \frac{h_3 - k_1 h_2}{h_3 - h_1 h_2}(h_2 + h_3 z^{-1})z^{-2}. \qquad (3.76)$$

Also in this case only one parameter can be selected for the highpass filter once the lowpass filter coefficients are chosen. Linear phase will not be possible in Case 2 as it would only lead to the trivial case with equal lowpass and highpass filters. A lattice filter structure is hardly possible here because the relation between the parameters in the lowpass and highpass filters is rather complex.

Case 3

Requirements:

1. $k_1 - h_1 = 0$

2. $k_3 - h_3 + h_2 k_1 - h_1 k_2 = 0$

With this requirement the highpass analysis filter becomes:

$$H_{HP}(z) = 1 + h_1 z^{-1} + k_2 z^{-2} + (h_3 + h_1(k_2 - h_2))z^{-3} \qquad (3.77)$$

The highpass filter has one free parameter, which in this case is selected as k_2. There exists two linear phase solutions for the analysis and synthesis filters. These solutions are

$$H_{LP}(z) = 1 \pm z^{-1} \pm z^{-2} + z^{-3}, \qquad (3.78)$$

and

$$H_{HP} = 1 \pm z^{-1} \mp z^{-2} - z^{-3} \qquad (3.79)$$

where the upper signs belong to one solution and the lower signs belong to
the other.

□

We now try to draw some general conclusions from the above example.
Originally, the filter polynomials have 6 parameters corresponding to 3 zeros
in each filter. When imposing the condition of PR and FIR synthesis filters,
corresponding to making the polyphase matrix determinant equal to a pure
delay, two conditions have to be satisfied. This is because the determinant
is of 2nd order. This leaves 4 free parameters for the system design. In the
example we chose the lowpass filter parameters to be free, which left only
one free parameter for the highpass filter. This is by no way necessary. The
free parameters can be freely distributed among the two filters. It would
also be possible to use some free parameters for the analysis filters and the
remaining parameters for the design of the synthesis filters.

How can this result be generalized? Assume that the two analysis filters
may have different orders, and denote the lowpass filter order by L and the
highpass filter order by K. That is, we altogether have $L + K$ parameters
before imposing the FIR PR condition.

Splitting in polyphase components and calculating the determinant, we
find that the determinant order is $\lfloor (L + K + 1)/2 \rfloor$. Setting all but one of
the determinant coefficients equal to 0, the coefficient relations reduce the
number of free parameters in the analysis system to $L + K + 1 - \lfloor (L + K + 1)/2 \rfloor$. The synthesis filters are, of course, completely determined from the
analysis filter coefficients.

We saw in the example above that there exist several solutions for PR
by requiring that all but one of the determinant coefficients be equal to
zero. If $L + K$ is an odd number, the situation is somewhat different. The
last coefficient in the determinant will consist of one filter coefficient or a
product of filter coefficients. To make this term equal to zero, one of the filter
coefficients has to be set to zero, thus reducing the filter order. Therefore,
the last coefficient in the determinant should be kept, and all the other
coefficients in the determinant polynomial be set to zero. The implication
is that for $L + K$ odd, only one PR solution exists.

Example

We illustrate this property further by using the simplest possible example. Let the filters be of lengths 3 and 2 for the lowpass and highpass filters, respectively. Going back to Equations 3.63 and 3.64, we obtain the present example by setting $h_3 = k_2 = k_3 = 0$. The determinant in Equation 3.66 reduces to:

$$\det \mathbf{P}(z) = (k_1 - h_1) + k_1 h_2 z^{-1}. \tag{3.80}$$

By setting the last term to zero, we must have either $k_1 = 0$ or $h_2 = 0$. In both cases this results in reduced filter order. So in this case, the only sensible PR solution requires that $k_1 = h_1$. The lowpass and highpass filters can still be made different through the free choice of h_2.

□

To sum up: There is exactly one type of PR condition when $L + K$ is odd. When $L + K$ is even, the number of different solutions is equal to the polyphase determinant order $\lfloor (L + K + 1)/2 \rfloor$.

3.4.5 Lattice Filters

In the previous section we examined simple two-channel PR systems, and were able to draw some general conclusions about possible filter lengths, the number of different solutions, and the number of free parameters. For the case of length four filters in both channels, we derived a lattice filter realization for the analysis and synthesis systems. These are possible due to a special relationship between the lowpass and highpass filters. We did not show that lattice structures exist for higher order filters, nor for filters with different lengths for lowpass and highpass filters. It is simple to derive lattice solutions when the lowpass and highpass filters are of lengths 4 and 2, respectively. In fact, this case produces two different degenerate lattice configurations. One solution is obtained by setting $d = 0$. The other solution is the limit when $d \to \infty$.

For length 6 filters it is tractable to derive the lattice relationship for one of the solutions from the general equations. For higher order it becomes cumbersome to find the lattice form due to the highly nonlinear relations between the coefficients.

In this section we show the opposite: For all lattice filters we prove that the perfect reconstruction property is guaranteed through the structure. It is plausible from the above example that the lattice structure does not generate all possible solutions. Still the lattices probably constitute some of the most important structures from an application point of view.

To arrive at higher order lattice systems, let us motivate the derivation by revisiting the structure of Figure 3.26. We have redrawn the same filter in Figure 3.29 but put in some extra boxes. There are two inputs and two

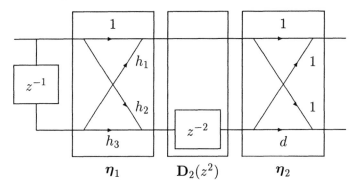

Figure 3.29: *Length 4 lattice analysis structure.*

outputs to each box. The box operations can be mathematically described by matrices as indicated in the figure. The matrices are:

$$\boldsymbol{\eta}_1 = \begin{bmatrix} 1 & h_1 \\ h_2 & h_3 \end{bmatrix}, \tag{3.81}$$

$$\mathbf{D}_2(z^2) = \begin{bmatrix} 1 & 0 \\ 0 & z^{-2} \end{bmatrix}, \tag{3.82}$$

$$\boldsymbol{\eta}_2 = \begin{bmatrix} 1 & 1 \\ 1 & d \end{bmatrix}. \tag{3.83}$$

The complete polyphase matrix is the product of the three matrices

$$\mathbf{P}(z^2) = \boldsymbol{\eta}_2 \mathbf{D}_2(z^2) \boldsymbol{\eta}_1. \tag{3.84}$$

The filter transfer vector is expressed by the polyphase matrix as

$$\mathbf{h}(z) = \mathbf{P}(z^2) \mathbf{d}_2(z). \tag{3.85}$$

In the following we only look at the properties of the polyphase matrix viewed as a function of z, i.e. $\mathbf{P}(z)$.

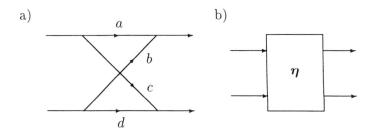

Figure 3.30: *a) Realization of a 2 × 2 lattice and b) its symbolic representation. η represents the lattice coefficients.*

The inverse of the polyphase matrix is given by

$$\mathbf{Q}(z) = \mathbf{P}^{-1}(z) = \boldsymbol{\eta}_1^{-1}\mathbf{D}_2^{-1}(z)\boldsymbol{\eta}_2^{-1}. \tag{3.86}$$

The inverse of the delay matrix $\mathbf{D}_2(z)$ represents a one sample *advance* in one branch, and is accordingly *noncausal*. We have mentioned earlier that this can be coped with by allowing an overall system delay. Formally this is expressed as

$$\mathbf{D}_2^{-1}(z) = z \left[\begin{array}{cc} z^{-1} & 0 \\ 0 & 1 \end{array} \right] = z\mathbf{D}_2^{(i)}(z). \tag{3.87}$$

In practice we use $\mathbf{D}_2^{(i)}(z)$ as the causal delay matrix inverse.

The main feature of the shown lattice structure is that the polyphase matrix has been factored in component matrices which have a simple inverse. There are two types of component matrices:

1. Delay-free coupling matrices.

2. Diagonal delay matrices.

The coupling matrix can be generalized to

$$\boldsymbol{\eta} = \left[\begin{array}{cc} a & b \\ c & d \end{array} \right], \tag{3.88}$$

which operate as a two-port and can be realized as in Figure 3.30 a. This building block is so important that we have introduced a special symbol for it, as given in Figure 3.30 b). The inverse is given by

$$\boldsymbol{\eta}^{-1} = \frac{1}{ad - bc} \left[\begin{array}{cc} d & -b \\ -c & a \end{array} \right], \tag{3.89}$$

and is depicted in Figure 3.31 a), with $k = 1/(ad - bc)$ and a symbolic block diagram in Figure 3.31 b).

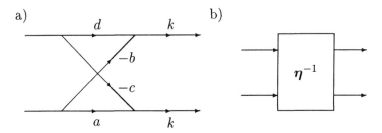

Figure 3.31: *a) Realization of a 2×2 reconstruction lattice and b) its symbolic representation.* η^{-1} *represents the lattice coefficients and the scaling factors.*

We can go one step further and normalize the coefficients in the analysis filters as

$$\eta = \left[\begin{array}{cc} \frac{a}{\sqrt{ad-bc}} & \frac{b}{\sqrt{ad-bc}} \\ \frac{c}{\sqrt{ad-bc}} & \frac{d}{\sqrt{ad-bc}} \end{array} \right], \tag{3.90}$$

then the coefficients in the synthesis filters become

$$\eta^{-1} = \left[\begin{array}{cc} \frac{d}{\sqrt{ad-bc}} & \frac{-b}{\sqrt{ad-bc}} \\ \frac{-c}{\sqrt{ad-bc}} & \frac{a}{\sqrt{ad-bc}} \end{array} \right]. \tag{3.91}$$

The system determinant is thus eliminated (equal to one) and the coefficients have the same numerical values in both filter banks.

It is obvious that we can generalize to higher order filters by cascading coupling lattices and delay matrices as

$$\mathbf{P}(z) = \boldsymbol{\eta}_L \, \mathbf{D}_2(z) \boldsymbol{\eta}_{L-1} \, \mathbf{D}_2(z) \cdots \mathbf{D}_2(z) \, \boldsymbol{\eta}_1. \tag{3.92}$$

The analysis polyphase lattice matrix system is shown in Figure 3.32. The polyphase matrix inverse is found as

$$\mathbf{Q}(z) = \boldsymbol{\eta}_1^{-1} \, \mathbf{D}_2^{(i)}(z) \cdots \mathbf{D}_2^{(i)}(z) \, \boldsymbol{\eta}_{L-1}^{-1} \, \mathbf{D}_2^{(i)}(z) \, \boldsymbol{\eta}_L^{-1}, \tag{3.93}$$

where the true delay operator inverse has been replaced by the causal pseudo-inverse. With this delay operator, the synthesis polyphase matrix has a structure as illustrated in Figure 3.33.

A comparison to the FIR QMF case is interesting. In QMF systems the highpass filter is directly given by the lowpass filter (see Equation 3.40).

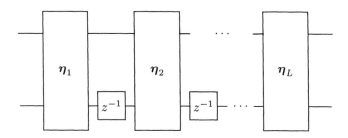

Figure 3.32: *Multistage two-channel lattice polyphase matrix.*

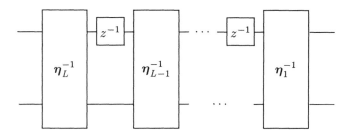

Figure 3.33: *Multistage two-channel polyphase synthesis matrix.*

Here, the highpass filter is left with one free parameter once the lowpass filter is specified. Thus we have some more free parameter in the lattice case in addition to the guaranteed PR situation.

Orthogonal Lattice Filters

Orthonormal filters forming unitary systems have enjoyed great popularity [3], partly because it has been believed that unitarity yields optimal performance in subband coders in the same way as unitary transforms are optimal in transform coders. We prove in Chapter 5 that nonunitary subband coder filter banks outperform the unitary filter banks, especially for systems with a small number of bands. Unitary systems may still be important because they may be easier to implement, and they are optimal when applying closed loop DPCM for subband quantization. (See Chapter 5).

The orthonormal case is obtained by putting constraints on the relation between the lattice coefficients as

$$\eta = \begin{bmatrix} \cos\phi & -\sin\phi \\ \sin\phi & \cos\phi \end{bmatrix}. \tag{3.94}$$

This lattice performs what is called a *Givens rotation* [50] and is lossless [3]. As the orthonormal structure is a special case of the previous theory, perfect reconstruction is no issue. However, it may be interesting to verify unitarity of the unit sample responses. A simple verification is based on the fact that the unit sample responses in the synthesis part are reversed versions of the analysis unit sample responses. A simple matrix property comes in handy: The product of unitary matrices is itself a unitary matrix. General unitary matrices can thus be constructed by cascading elementary unitary matrices.

If we relax the orthonormality requirement to only orthogonality, we obtain a simpler implementation. By dividing all coefficients by $\cos \phi$, we get

$$\boldsymbol{\eta} = \begin{bmatrix} 1 & -\tan \phi \\ \tan \phi & 1 \end{bmatrix} = \begin{bmatrix} 1 & -k \\ k & 1 \end{bmatrix}. \tag{3.95}$$

The inverse is now

$$\boldsymbol{\eta}^{-1} = \frac{1}{1 + \tan^2(\phi)} \begin{bmatrix} 1 & \tan \phi \\ -\tan \phi & 1 \end{bmatrix} = -\cos^2(\phi)\boldsymbol{\eta}^T. \tag{3.96}$$

The scaling factors of the type $\cos^2(\phi)$ can be collected in a scaling factor at the system output, thus adding little to the system complexity. Each lattice uses two multiplications, only in this orthogonal system. In contrast, the orthonormal case needs four multiplications per lattice.

Linear Phase Lattice Structures

Another special variety of the general lattice theory is the linear phase case. Linear phase systems are obtained by choosing the lattice component matrix as

$$\boldsymbol{\eta} = \begin{bmatrix} 1 & \tan \phi \\ \tan \phi & 1 \end{bmatrix} = \begin{bmatrix} 1 & k \\ k & 1 \end{bmatrix}, \tag{3.97}$$

as will be shown below. The inverse lattice is here given by

$$\boldsymbol{\eta}^{-1} = \frac{1}{1 - k^2} \begin{bmatrix} 1 & -k \\ -k & 1 \end{bmatrix}. \tag{3.98}$$

We see that the inverse lattice has the same structure, except for the multiplier sign and the normalization factor given by the matrix inverse.

The two unit sample responses from the analysis system input to any output of the coupling matrices, $\boldsymbol{\eta}$, are the reverses of each other. The z-domain equivalent is that the transfer polynomials are inversed versions of each other. We can prove this by induction. Let us first look at the output from the first coupling lattice in the analysis filter system:

$$\mathbf{h}_1(z) \;=\; \left[\begin{array}{c} r_1(z) \\ r_2(z) \end{array}\right] = \left[\begin{array}{cc} 1 & k_1 \\ k_1 & 1 \end{array}\right] \mathbf{d}_2(z) \tag{3.99}$$

$$= \left[\begin{array}{cc} 1 & k_1 \\ k_1 & 1 \end{array}\right]\left[\begin{array}{c} 1 \\ z^{-1} \end{array}\right] = \left[\begin{array}{c} 1 + k_1 z^{-1} \\ k_1 + z^{-1} \end{array}\right] = \left[\begin{array}{c} r_1(z) \\ f_1(z) \end{array}\right] . \tag{3.100}$$

As we can see, the two obtained polynomials are reverses of each other. Assume that we after i stages also have resulting polynomials called $f_i(z)$ in one branch and $r_i(z)$ in the other branch which are reverses of each other, that is

$$r_i(z) = f_i(z^{-1})z^{-(2i-1)}. \tag{3.101}$$

After the next stage the transfer polynomials then become

$$\mathbf{h}_{i+1}(z) \;=\; \left[\begin{array}{c} r_{i+1}(z) \\ f_{i+1}(z) \end{array}\right] = \left[\begin{array}{cc} 1 & k_{i+1} \\ k_{i+1} & 1 \end{array}\right]\left[\begin{array}{cc} 1 & 0 \\ 0 & z^{-2} \end{array}\right]\left[\begin{array}{c} r_i(z) \\ f_i(z) \end{array}\right]$$

$$= \left[\begin{array}{c} f_i(z^{-1})z^{-(2i-1)} + k_{i+1}f_i(z)z^{-2} \\ k_{i+1}f_i(z^{-1})z^{-(2i-1)} + f_i(z)z^{-2} \end{array}\right]$$

$$= \left[\begin{array}{c} \{f_i(z^{-1})z^2 + k_{i+1}f_i(z)z^{2i-1}\}\, z^{-(2(i+1)-1)} \\ f_i(z)z^{-2} + k_{i+1}f_i(z^{-1})z^{-(2i-1)} \end{array}\right] . \tag{3.102}$$

The polynomials $f_{i+1}(z)$ and $r_{i+1}(z)$ are still reverse of each other, which completes the proof.

A pair of linear phase responses can be obtained from $r_i(z)$ and $f_i(z)$ by delaying one with respect to the other by an arbitrary number of samples, K, and then adding and subtracting them. This is the same as including a matrix $\mathbf{D}_2(z^K)$ in the polyphase matrix. It is, however, advantageous to choose the sample delays as $K = 2k$, $k \in Z_+$ to facilitate the use of the noble identities for the downsamplers in the analysis filter bank.

The analysis filter bank is thus given by

$$\mathbf{h}(z) = \mathbf{P}(z^2)\mathbf{d}_2(z) = \mathbf{H}_2 \mathbf{D}_2(z^{2k})\boldsymbol{\eta}_L\, \mathbf{D}_2(z^2)\boldsymbol{\eta}_{L-1}\, \mathbf{D}_2(z^2)\cdots \mathbf{D}_2(z^2)\,\boldsymbol{\eta}_1 \mathbf{d}_2(z). \tag{3.103}$$

We have chosen the special matrix $\mathbf{H}_2$, the 2×2 Hadamard matrix, as the final operator of the polyphase matrix. By varying k, different filters are found which have different length with the same number of multiplications.

Filter Example

Assume we have a linear phase lattice filter given by

$$\mathbf{h}(z) = \mathbf{H}_2 \mathbf{D}_2(z^{2k}) \boldsymbol{\eta}_2 \mathbf{D}_2(z^2) \boldsymbol{\eta}_1 \mathbf{d}_2(z), \tag{3.104}$$

where each lattice matrix $\boldsymbol{\eta}_i$ is described by the single parameter k_i. The transfer vector from the input of the analysis filter, given the above polyphase matrix, to the input to the final delay matrix is

$$\begin{bmatrix} r_2(z) \\ f_2(z) \end{bmatrix} = \boldsymbol{\eta}_2 \mathbf{D}_2(z^2) \boldsymbol{\eta}_1 \mathbf{d}_2(z) = \begin{bmatrix} 1 + k_1 z^{-1} + k_1 k_2 z^{-2} + k_2 z^{-3} \\ k_2 + k_1 k_2 z^{-1} + k_1 z^{-2} + z^{-3} \end{bmatrix}.$$
$$\tag{3.105}$$

where $r_2(z)$ represents the upper branch, and $f_2(z)$ represents the lower branch. We can now choose the delay k in the matrix $\mathbf{D}_2(z^{2k})$. We consider two cases:

Case 1: $k = 2$.

The transfer functions are:

$$H_{LP/HP}(z) = 1 + k_1 z^{-1} + (k_1 \pm 1)k_2 z^{-2} \pm (k_1 \pm 1)k_2 z^{-3} \pm k_1 z^{-4} \pm z^{-5}. \tag{3.106}$$

Case 2: $k = 4$.

The transfer functions are:

$$H_{LP/HP}(z) = 1 + k_1 z^{-1} + k_1 k_2 z^{-2} + k_2 z^{-3} \pm k_2 z^{-4} \pm k_1 k_2 z^{-5} \pm k_1 z^{-6} \pm z^{-7}. \tag{3.107}$$

In both cases the lowpass filter is symmetric, while the highpass filter is antisymmetric. The filter order is different in the two cases, but the lattice coefficients are the same.

The two derived filters have been optimized for *coding gain* (see Chapter 5) for an AR(1) process with correlation coefficient $\rho = 0.95$. Figure 3.34

shows the magnitude responses of the length 6 filters in Equation 3.106, and Figure 3.35 shows the corresponding plots for the length 8 filters in Equation 3.107.

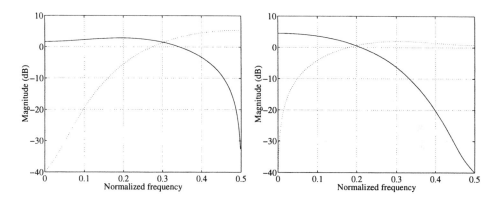

Figure 3.34: *Magnitude responses of analysis filters (left) and synthesis filters (right) of the optimized linear phase filter in Equation 3.106. Lattice coefficients:* $k_1 = 6.8328$, $k_2 = -4.5627$ *in the first and second stage of the analysis filter. Coding gain=6.21 dB.*

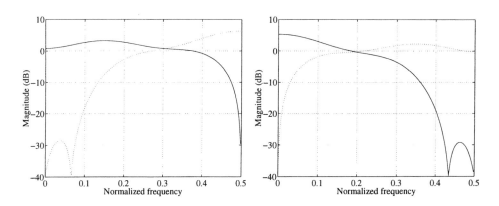

Figure 3.35: *Magnitude responses of analysis filters (left) and synthesis filters (right) of the optimized linear phase filter in Equation 3.107. Lattice coefficients:* $k_1 = -0.2187$, $k_2 = -10.2459$ *in the first and second stage of the analysis filter. Coding gain=6.38 dB.*

The characteristic forms of the filters can be understood from the optimization criteria in Chapter 5. It is interesting to note that the coding gain obtained for the longest filter is higher by 0.17 dB. One might conclude that the filter length is important in addition to the number of free parameters.

□

3.5   Tree Structured Filter Banks

There are, in principle, two basic ways of constructing N-channel filter banks. The most obvious solution is to use bandpass filters in parallel. This gives a large flexibility: nonuniform frequency partitioning is possible as well as individual selection of filter orders. In general, the complexity becomes high. It is also possible within the framework of parallel filter banks to impose certain dependencies among the channel coefficients to reduce the complexity, as for example in cosine modulated filter banks, as will be discussed in the next main section. The second family of filter structures are the tree-structured filter banks. In this case the signal is first split in two, or any other small number of channels. The resulting outputs are input to a second stage with further separation. This process can go on as indicated in Figure 3.36 a) for a system where at every stage the outputs are split further until the required resolution has been obtained. In Figure 3.36 b) the synthesis filter bank is illustrated.

Tree-structured systems have a rather high flexibility. Nonuniform filter banks are obtained by splitting only some of the outputs at each stage. Different filters can be applied at the different stages to adapt to varying requirements for different channels. To guarantee perfect reconstruction, the synthesis filter bank must at each stage reconstruct the input signal to the corresponding analysis filter.

Tree structured filter banks also have fairly low complexity. They are quite simple to design because each analysis/synthesis filter pair can be designed separately. Optimization, however, with respect to gain, blocking effect reduction, etc. becomes more involved for tree-structured filter banks. Gain optimization requires the knowledge of the input spectrum at each stage. If the input signal spectrum is simple, the input spectrum to the next stages are modified through filtering, and may be quite complex.

The tree-structured system based on two-channel subsystems is flexible

a)   b)

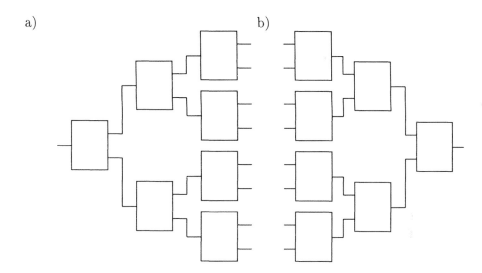

Figure 3.36: *a) Tree structured analysis filter bank where at each stage the signal is split in two and decimated by a factor of two to obtain critical sampling. b) Corresponding synthesis filter bank for recombination and interpolation of the signals.*

for making adaptive systems. The tree-growing can be changed during application of the filter bank to make a tradeoff between frequency and time resolution adaptive to the image content. An illustration of this idea is shown in Figure 3.37.

3.6 Parallel Filter Banks

Parallel filter banks come in different flavors. The conceptually simplest form is composed of a bank of bandpass filters followed by appropriate downsampling. If applying these filters the PR condition can only be satisfied through optimization of the filter coefficients. Such filter banks tend to have high computational complexity. The second main form has some type of inherent structure, that is, there exists some coupling between the channels which can facilitate 1) the PR condition and 2) lower the computational complexity. The polyphase structure, which was introduced in Section 3.3, can represent general coupled forms if the polyphase matrix $\mathbf{P}(z)$ is selected favorably. A

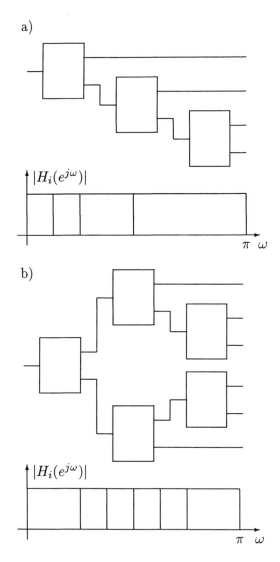

Figure 3.37: *Two examples of tree-growing in a tree-structured analysis filter bank, and their respective idealized bandpass characteristics.*

well-known example of such a system is a filter bank solely based on a DFT, which for efficiency is implemented using some form of FFT algorithm. In this case $\mathbf{P}(z)$ is the DFT matrix. It turns out that an important class of efficient subband filter banks can be derived from the DFT and related transforms. These can be classified as *modulated filter banks* because the bandpass filters are derived from a lowpass prototype filter either by real or complex modulation. We will develop practical modulated filter banks based on both FIR or IIR filters. Additionally, we present more general systems where the polyphase matrix consists of easily invertible matrices in cascade. The generalization from modulated filter banks turns out to be quite important for eliminating visual artifacts in subband coders due to a higher degree of freedom in filter optimization.

3.6.1 Allpass-based IIR Parallel Systems

A two-channel filter bank based on allpass filters was discussed in Section 3.4.3. It was found that allpass filters have some unique properties that make them very interesting for constructing IIR systems with the perfect reconstruction property, provided we use anticausal filters either in the analysis or the synthesis filter bank. The structure offers minimal processing as well as exceptionally good coefficient sensitivity and noise performance. It would therefore be nice if these important features could be carried over to a multichannel case with allpass filters and some applicable transform. Hopefully, such constructions will offer even lower complexity than tree-structured systems based on the two-channel building blocks.

The transfer vector of the two-channel analysis system is given by

$$
\mathbf{h}_2(z) = \begin{bmatrix} 1 & 1 \\ 1 & -1 \end{bmatrix} \begin{bmatrix} A_0(z^2) & 0 \\ 0 & A_1(z^2) \end{bmatrix} \begin{bmatrix} 1 \\ z^{-1} \end{bmatrix} = \mathbf{H}_2 \mathbf{A}_2(z^2) \mathbf{d}_2(z).
$$

(3.108)

$\mathbf{H}_2$ is a 2×2 *Hadamard transform* which is responsible for the lowpass-highpass modulation.

When generalizing to N channels, we make a straight forward extension of the above equation. However, the Hadamard transform is not appropriate any longer. In fact, the matrix $\mathbf{H}_2$ can also represent an inverse, unscaled *discrete Fourier transform* (DFT). For N channels this transform is the appropriate modulation transform, as will be demonstrated shortly.

The N-channel transfer matrix is then given by

$$\mathbf{h}_N(z) = \mathbf{W}_N^{-1}\mathbf{A}(z^N)\mathbf{d}_N(z). \tag{3.109}$$

In this equation $\mathbf{A}(z) = \mathrm{diag}\,[A_0(z), A_1(z), \dots A_{N-1}(z)]$ is a matrix with diagonal elements consisting of allpass filters, $\mathbf{W}_N^{-1}$ is the unscaled IDFT matrix with elements $\exp(j\frac{2\pi}{N}kn)$, and $\mathbf{d}_N(z) = [1, z^{-1}, \dots, z^{-(N-1)}]^T$ is a vector of delays. The polyphase analysis matrix is thus composed of the two terms in Equation 3.109:

$$\mathbf{P}(z) = \mathbf{W}_N^{-1}\mathbf{A}(z) \tag{3.110}$$

The resulting filter bank, after applying the noble identities (see Appendix A), is shown in Figure 3.38.

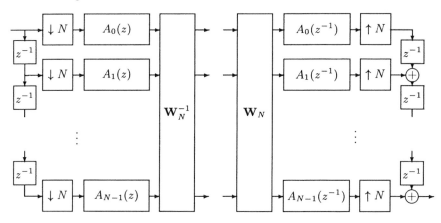

Figure 3.38: *Left: Analysis filter bank based on allpass filters and a DFT. Right: Possible inverse system*

Because the system can be written in the standard form described in Section 3.4, the inverse filter bank has the general polyphase synthesis structure with polyphase matrix given as

$$\mathbf{Q}(z) = \mathbf{P}^{-1}(z) = \mathbf{A}^{-1}(z)\mathbf{W}_N = \mathbf{A}(z^{-1})\mathbf{W}_N. \tag{3.111}$$

We obtain the last equality because the inverse of a diagonal matrix is diagonal with elements equal to the inverse of each element and for allpass filters $A_{k,k}^{-1}(z) = A_{k,k}(z^{-1})$ (see Appendix A). The polyphase synthesis structure can thus be implemented as given in Figure 3.38.

Alternatively, we can verify the PR property of the combined analysis and synthesis filters through the following observations:

- The overall I/O transfer function does not change if we eliminate the IDFT and the DFT which are directly cascaded and are inverse transforms.

- After removing the transforms, the allpass filters in the analysis and synthesis systems are cascaded. Because $A_i(z)A_i(z^{-1}) = 1$, $i \in 0, 1, \ldots, N-1$ from the allpass filter property, all allpass filters can be replaced by short circuits without changing the overall I/O transfer function.

- After eliminating the allpass filters, the remaining system consists of a multiplexer and a demultiplexer. They are inverse systems which leave the samples unchanged except for the delay.

As in the two-channel case a stable implementation of $A_i(z^{-1})$ depends on the use of an anticausal system. Anticausality can be realized for processing finite length signals by reversing the input and output sequences and replacing the anticausal filter with its causal counterpart as described in Appendix A.

Having proven the PR condition and verified that the system can be implemented using stable filters, it remains to be shown that the system can provide meaningful filtering. This is done next.

We first construct a prototype filter consisting of polyphase allpass components according to

$$H_0(z) = \sum_{k=0}^{N-1} A_k(z^N)z^{-k}. \tag{3.112}$$

This filter corresponds to the first entry in $\mathbf{h}(z)$. It is not obvious that the above expression will constitute flexible and meaningful prototype filters. In Section 3.4.3 we argued that this type of filter construction does make a very efficient lowpass or highpass filter for the two-channel case $(N = 2)$. Renfors and Saramäki [44] introduced a Remez-type algorithm for the design of lowpass filters conforming to Equation 3.112 for arbitrary N. In fact, as in the two-channel case, the filter structure yields very efficient filters in terms of computational complexity, coefficient sensitivity, and noise.

From the prototype filter we derive bandpass filters through complex modulation, that is a bandpass filter $H_n(z)$ with center frequency at $\omega_n = 2\pi n/N$ can be written as

$$H_n(z) = H_0(ze^{-j\frac{2\pi}{N}n}) = \sum_{k=0}^{N-1} A_k(z^N)e^{j\frac{2\pi}{N}kn}z^{-k}. \qquad (3.113)$$

Using this expression for the channel filters of the generic analysis filter bank of Figure 3.2, we can express the subband signals, prior to the subsampling operation, as

$$Y_n'(z) = H_n(z)X(z) = \sum_{k=0}^{N-1}\{z^{-k}A_k(z^N)X(z)\}e^{j\frac{2\pi}{N}kn}, \quad n = 0, 1, \cdots, N-1.$$
$$(3.114)$$

This can be interpreted as the IDFT (less a scaling factor) of the signal entity composed of one output signal sample from each of the filters $z^{-k}A_k(z^N)$, $k = 0, 1, \ldots, N-1$. When the $Y_n'(z)$s are subsampled, the noble identities of Appendix A can be applied. The resulting structure producing the subband signals are as shown on the left hand side of Figure 3.38.

It is important to observe the channel arrangement in this type of filter bank. Figure 3.39 shows the frequency regions of each filter.

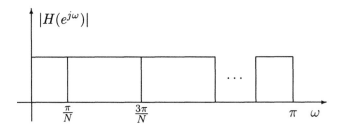

Figure 3.39: *Ideal frequency partitioning in the complex modulated filter bank.*

There are only N channels that cover the frequency range $\omega \in [0, 2\pi]$, but the frequencies above $\omega = \pi$ contain the same information as the frequencies below that frequency if the input signal is real. However, for N even, $N/2-1$ of the channels are complex. After decimation by N, the real and imaginary parts of each channel will be critically sampled and thus have a one-sided bandwidth of π/N. These will altogether represent $N-2$ channels. The two remaining channels, the lowpass and highpass channels will after decimation remain real and therefore add up to the N required channels.

We have now shown that the filter bank consists of modulated filters derived from a prototype filter. The prototype filter, composed of a sum of allpass filters, is designed to act as a lowpass filter. We have also demonstrated that the analysis-synthesis filters constitute a PR system provided one of the filter banks is operated in an anticausal mode.

Example

The simplest nontrivial filter bank consists of

$$A_0(z^N) = 1, \tag{3.115}$$

in one branch and first order allpass filters

$$A_i(z^N) = \frac{\alpha_i - z^{-1}}{1 - \alpha_i z^{-1}}, \quad i \in 1, 2, \cdots, N-1, \tag{3.116}$$

in the other branches.

Frequency response examples for $N = 8$ are shown in Figure 3.40. Observe the stopband maxima at $m\pi/8$ for $m = 3, 5, 7$. These maxima are unavoidable, but for a given filter order, as in our example, there is a tradeoff between the width of the maxima and the stopband attenuation. Actually, the passband ripple also depends on the stopband attenuation because the polyphase components are allpass filters. For our application the passband ripple will be of no concern for practical stopband attenuations. If we increase the subfilter orders, we still get the stopband maxima at the same frequencies and with the same maximum values, but the peaks get narrower if we maintain the same stopband attenuation, or we can keep the original peaks and obtain more stopband attenuation elsewhere.

Another important property of the allpass-based DFT-modulated analysis filter bank is that the output signals are complex, except for the first and last channels. For the eight-band system there are two real channels and three complex channels, which add up to eight channels counting the real and imaginary parts as independent channels. Coding examples using this type of filter bank can be found in [51].

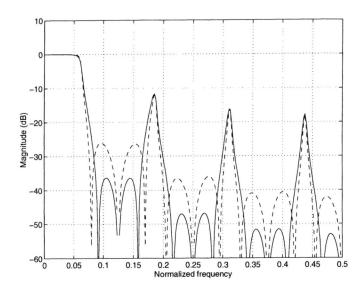

Figure 3.40: *Magnitude responses of two lowpass prototype filters consisting of first-order allpass sections in parallel for an eight-channel case.*

Summary of the DFT-Modulated Filter Bank based on Allpass Filters.

The filter bank has the following properties:

- It constitutes a PR system provided that anticausal processing can be applied either in the analysis or synthesis filter bank.

- It produces $N/2 - 1$ complex channels and two real channels.

- The stopband contains $N/2 - 1$ spectral peaks with maxima independent of the filter parameters for the given no. of channels.

- The filter bank has very low complexity if implemented with efficient allpass filter structures and optimal FFT algorithms.

- The system has very low coefficient sensitivity due to structural constraints similar to those of the two-channel system. This also indicates good quantization noise property.

3.6.2 Cosine Modulated FIR Filter Banks

The previous section presented a very efficient IIR filter bank which splits the signal into complex channels resulting in a minimum (Nyquist) representation. In this section we will consider the possibility of constructing real filter banks with a somewhat similar structure, but this time based on FIR filters. The resulting systems are denoted *cosine-modulated filter banks* because all the bandpass filters are cosine-modulated versions of prototype lowpass FIR filters. With a uniform channel separation, an efficient implementation consisting of a DFT-related transform and subfilters with critical sampling is possible.

Although FIR filter banks are not expected to compete with IIR filter banks with regard to computational complexity, there are issues such as perfect reconstruction of infinite length signals and stability, which still may favor FIR-based filter banks. Another advantage is the better design flexibility offered by FIR filters.

The cosine-modulated filter banks were originally derived in the transmultiplexer literature [39]. However, transmultiplexers were developed without requiring perfect reconstruction in the analysis-synthesis filter bank combination. An early attempt to use this type of construction in coding was made in [24], where an efficient analysis filter bank implementation was developed using simple mathematical manipulations. Because methods for aliasing cancelation were not known, the design was made as a compromise between severe aliasing and severe spectral gaps. This is like choosing between two evils.

Malvar's LOT [52] is the earliest known PR cosine-modulated filter bank, but the first LOT version was limited to filters where each filter length had to be equal to $2N$ for N channels. Filter banks without this length limitation were thereafter simultaneously presented by three independent authors [33, 34, 35].

The Analysis Filter Bank

We limit our discussion to uniform filter banks where each of the N channels occupies the same bandwidth, namely π/N, and the channel filter band edges are ideally located at $n\pi/N$, for $n = 0, 1, ..., N$. The channel arrangement is shown in Figure 3.41.

A prototype lowpass filter of length $L = 2Nq$, $q \in Z_+$ is defined by its

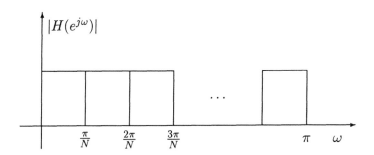

Figure 3.41: *Channel arrangement of a cosine-modulated filter bank*

unit sample response $h(k)$. The unit sample response of channel no. n can be obtained by modulating the prototype filter unit sample response as

$$h_n(k) = h(k) \cos\left(\frac{\pi}{N}(n + \frac{1}{2})(k + \frac{1}{2})\right) = h(k)T_N(n, k). \qquad (3.117)$$

The modulation term $T_N(n, k)$ represents elements of a matrix $\mathbf{T}_N$, the cosine transform of type IV [38], originally called a *double odd Fourier transform* [53]. The index N indicates the no. of channels. The matrix size will be discussed below. This transform has been chosen among several alternatives because it will provide the correct channel arrangement and has favorable symmetry properties. Other transforms will also do the job [3], but we limit our discussion to this particular case.

In the transform $n + 1/2$ accounts for a frequency shift of the prototype filter where n is the channel number, and the second parameter $1/2$ is inserted to make the transform completely symmetric.

We state the following two useful properties of this transform that can be easily verified by simple trigonometric identities:

1. Periodic extension:

$$T_N(n, m + 2Nr) = (-1)^r T_N(n, m) \qquad (3.118)$$

2. Symmetry:

$$T_N(n, 2N - 1 - m) = -T_N(n, m) \qquad (3.119)$$

Using the discrete convolution formalism, we can calculate the output signals in channel no. n as

$$y_n(k) = \sum_{l=0}^{L-1} x(k-l)h(l)T_N(n,l), \quad n \in 0,1,\ldots,N-1. \tag{3.120}$$

To derive an efficient structure with subfilters and an explicit transform, the summation is split into a double sum by introducing two summation variables instead of l as $l = m + 2Nr$, where $m \in 0,1,\ldots,2N-1$ and $r \in 0,1,\ldots,q-1$. In the following expression we also downsample the output signals by a factor N, that is we replace k by iN, to obtain critical sampling. This implies that we conserve the total number of samples in the filtering process. After some simple manipulations, we obtain

$$y_n(iN) = \sum_{m=0}^{2N-1} T_N(n,m)v_m(i), \tag{3.121}$$

where the input signal to the transform is given by

$$v_m(i) = \sum_{r=0}^{q-1} \kappa_m(r)u_m(i-2r). \tag{3.122}$$

The above equation represents convolution in subfilters using coefficients related to the prototype filter coefficients through

$$\kappa_m(r) = (-1)^r h(m + 2Nr). \tag{3.123}$$

Observe that the subfilter coefficients are subsampled (by $2N$) versions of the lowpass prototype filter modified by $(-1)^r$. The factor $(-1)^r$ originates from the periodic extension of the index in the transform matrix. The inputs to the subfilters are selected samples of the input signal

$$u_m(i) = x(Ni - m). \tag{3.124}$$

If transformed to the z-domain, the filtering operations in Equation 3.122 are expressed as:

$$V_m(z) = K_m(z^2)U_m(z), \tag{3.125}$$

where signal relations are given by $V_m(z) = Z\{v_m(i)\}$, and $U_m(z) = Z\{u_m(i)\}$, and the subfilter transfer functions are expressed as

$$K_m(z^2) = \sum_{r=0}^{q-1} \kappa_m(r)z^{-2r}. \tag{3.126}$$

The significance of the argument z^2 in the last equation is that there are two delays between each nonzero coefficient in the filters. A block diagram illustrating the operations in the above equations is given in Figure 3.42.

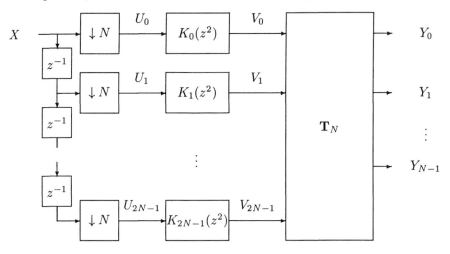

Figure 3.42: *Analysis filter bank structure*

The filter system looks like a polyphase structure. But in this case there are $2N$ subfilters, the channels are downsampled by N, and there are N independent output channels. Another peculiarity is the z^2-dependency of the subfilters.

We can modify the obtained system by observing certain relations between the different inputs to the subfilters and the symmetry of the transform. This simplification will

- facilitate the derivation of the synthesis filter bank,

- simplify the implementation,

- help us find a lattice structure with inherent PR.

The first modification is based on the relation

$$u_{m+N}(i) = u_m(i - 1). \tag{3.127}$$

The input multiplexer in Figure 3.42 can therefore be replaced by the input multiplexer in Figure 3.43. The second modification is based on the symmetry property of the transform $\mathbf{T}_N$:

$$y_n(i) = \sum_{m=0}^{2N-1} T_N(n,m)v_m(i) = \sum_{m=0}^{N-1} T_N(n,m)[v_m(i) - v_{2N-1-m}(i)]. \quad (3.128)$$

With only N inputs the transform is square of size $N \times N$. With the two given modifications above, the filter bank can be implemented as given in Figure 3.43.

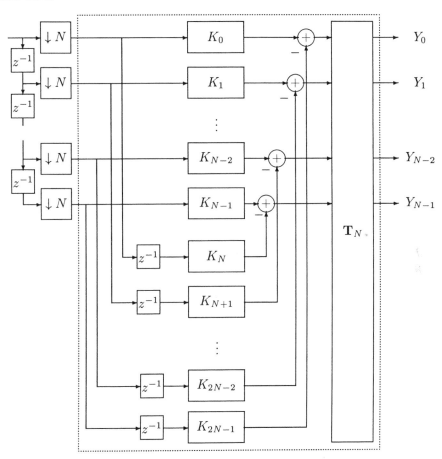

Figure 3.43: *Modified version of the cosine-modulated analysis filter bank. The inside of the dotted box represents the polyphase matrix.*

The polyphase matrix of the cosine-modulated filter bank can be expressed as the product of the cosine transform and the matrix representing the filter operations

$$\mathbf{P}(z) = \mathbf{T}_N \mathbf{K}(z). \tag{3.129}$$

The structure of the $N \times N$ matrix $\mathbf{K}(z)$ is very important. It consists of elements only along the two main diagonals:

$$\mathbf{K}(z) = \begin{bmatrix} K_0 & 0 & \cdots & 0 & -z^{-1}K_{2N-1} \\ 0 & K_1 & \cdots & -z^{-1}K_{2N-2} & 0 \\ \vdots & \vdots & \vdots & \vdots & \vdots \\ 0 & -z^{-1}K_{N+1} & \cdots & K_{N-2} & 0 \\ -z^{-1}K_N & 0 & \cdots & 0 & K_{N-1} \end{bmatrix}. \tag{3.130}$$

The argument z^2 is left out for all K's in the matrix elements. Notice the negative sign in front of all the antidiagonal elements originating from the minus-sign in Equation 3.128. Observe also that the polyphase matrix is a function of z, not only of z^2.

From the matrix $\dot{\mathbf{K}}(z)$ we extract simple submatrices $\mathbf{M}_i(z)$, consisting of four nonzero elements located symmetrically around the center of the matrix:

$$\mathbf{M}_i(z) = \begin{bmatrix} K_i(z^2) & -z^{-1}K_{2N-1-i}(z^2) \\ -z^{-1}K_{N+i}(z^2) & K_{N-1-i}(z^2) \end{bmatrix}, \quad i \in 0, 1, \ldots, N/2 - 1 \tag{3.131}$$

The submatrices represent separate two-ports that can be directly found from Figure 3.43. In this figure we see that there is always interconnection between two and two branches, only. The submatrices $\mathbf{M}_i(z)$ are called *superlattices* and one such matrix is shown in Figure 3.44.

We are now ready to derive the synthesis filter bank and the conditions for perfect reconstruction, which in a special case can be guaranteed structurally.

Synthesis Filter Bank in the Cosine-Modulated System

With all the preparations in the previous section, it is straight forward to find the formal inverse of the system. The polyphase matrix inverse is given by

$$\mathbf{Q}(z) = \mathbf{P}^{-1}(z) = \mathbf{K}^{-1}(z)\mathbf{T}_N^{-1} = \mathbf{K}^{-1}(z)\frac{2}{N}\mathbf{T}_N. \tag{3.132}$$

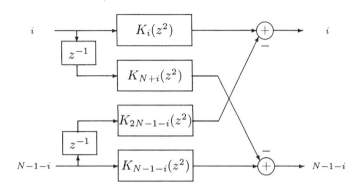

Figure 3.44: *Superlattice in the cosine-modulated analysis bank*

The last equality is obtained because the cosine transform matrix is its own inverse, except for a scaling factor $2/N$. The inverse $\mathbf{K}^{-1}(z)$ can be constructed from the inverses of $\mathbf{M}_i(z)$ in Equation 3.131 as

$$\mathbf{M}_i^{-1}(z) = \frac{1}{\det \mathbf{M}_i(z)} \begin{bmatrix} K_{N-1-i}(z^2) & z^{-1}K_{2N-1-i}(z^2) \\ z^{-1}K_{N+i}(z^2) & K_i(z^2) \end{bmatrix}, \qquad (3.133)$$

where

$$\det \mathbf{M}_i(z) = K_i(z^2)K_{N-1-i}(z^2) - z^{-2}K_{N+i}(z^2)K_{2N-1-i}(z^2). \qquad (3.134)$$

Based on these expressions we find the synthesis superlattice as illustrated in Figure 3.45.

The complete $\mathbf{K}^{-1}(z)$ can be found by putting the elements of each submatrix $\mathbf{M}_i^{-1}(z)$ back into this matrix in the position where the elements of $\mathbf{M}_i(z)$ where picked from $\mathbf{K}(z)$.

If the analysis filters are of the FIR type, the determinant expression of the synthesis superlattices will, in general, represent all-pole filters. These are possibly unstable. In such cases we have to split the filter in a causal part (with poles inside the unit circle) and an anticausal part (with poles outside the unit circle). The causal part is used in a standard way, while the anticausal part is treated as suggested in the Appendix A. Because the all-pole filters represent a significant part of the system complexity, it is preferable to eliminate them altogether, as we shall find convenient in the

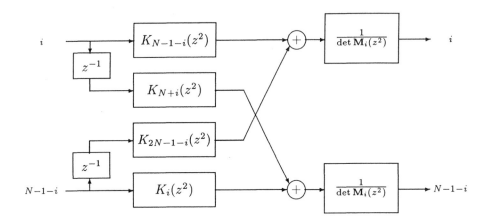

Figure 3.45: *Structure of a two-pair synthesis superlattice including the appropriate determinant.*

sequel. The determinant is required to be equal to a constant times a delay term, which is equivalent to eliminating all but one of the coefficients in the polynomial $\det \mathbf{M}_i(z)$.

Having obtained the inverse of the polyphase matrix, we can draw the complete synthesis filter bank structure in Figure 3.46, which looks pretty much like a transposed version of the analysis filter bank structure in Figure 3.42.

Simple PR System

In one particular case, the PR condition can easily be met such that the superlattice determinant is made equal to a pure delay. This is the case when each subfilter has only one nonzero coefficient.

Due to the delays in the superlattices, the simplest applicable form of Equation 3.131 is

$$\mathbf{M}_i(z) = \boldsymbol{\mu}_i(z) = \left[\begin{array}{cc} a_i z^{-2} & b_i z^{-1} \\ c_i z^{-1} & d_i \end{array} \right]. \tag{3.135}$$

With the two-sample delay attached to the coefficient a_i, the matrix inverse is given by

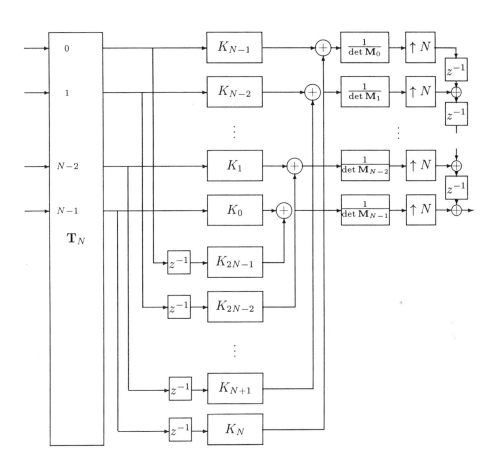

Figure 3.46: *Synthesis filter bank in a cosine-modulated system. The subfilters are equal the analysis subfilters, only located differently. The synthesis filter bank also includes the superlattice inverse determinants.*

$$\mathbf{M}_i^{-1}(z) = \boldsymbol{\mu}_i^{-1}(z) = \frac{1}{(a_i d_i - b_i c_i)z^{-2}} \begin{bmatrix} d_i & -b_i z^{-1} \\ -c_i z^{-1} & a_i z^{-2} \end{bmatrix}. \qquad (3.136)$$

The determinant consists of z^{-2}, which in the filter represents a two-sample advance, and a constant. When allowing an extra delay of two samples, z^{-2} can be dropped. The lattice realizations of these two equations are shown in Figure 3.47.

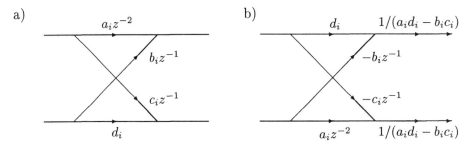

Figure 3.47: *Illustrations of the simple lattice structures. a) Analysis two-port. b) Synthesis two-port.*

Leaving out the downsamplers, the unit sample response lengths of this filter bank is $L = 2N$. Using this particular filter length the prototype filter can be designed by standard design methods. The lattice coefficients are directly related to the unit sample response of the prototype filter. For higher order systems there is, in general, no simple relation between the prototype filter and the lattice coefficients. It is therefore necessary to optimize the filters using the lattice coefficients as variables.

Example

In this example there are 8 channels and filter lengths are accordingly 16 samples. The prototype filter is designed using the Remez algorithm. Denote the derived unit sample response of the filter by $[h(0), h(1), \ldots, h(15)]$. From the previous theory we can find the coefficients in the lattice structure expressed by this unit sample response. Table 3.1 shows the relation explicitly. The index i point to the superlattice number.

An example of frequency magnitude responses for this eight-channel filter bank is illustrated in Figure 3.48. In both filter banks the magnitudes are

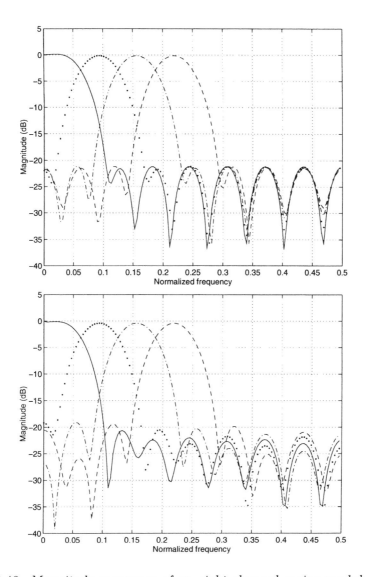

Figure 3.48: *Magnitude responses of an eight-channel cosine-modulated system with filters lengths $L = 16$. Only the four first channels are shown. The four upper channel responses are mirror images of the four lower. Top curves: Analysis filter bank responses. Lower curves: Synthesis filter bank responses.*

i=	$a_i =$	$b_i =$	$c_i =$	$d_i =$
1	$-h(12)$	$-h(11)$	$-h(4)$	$h(3)$
2	$-h(13)$	$-h(10)$	$-h(5)$	$h(2)$
3	$-h(14)$	$-h(9)$	$-h(6)$	$h(1)$
4	$-h(15)$	$-h(8)$	$-h(7)$	$h(0)$

Table 3.1: *Relation between the prototype filter unit sample response and the lattice coefficients.*

scaled to approximately 0 dB in the passbands to make the comparison simpler.

□

In the above example the magnitude responses are fairly similar on the analysis and synthesis sides. They may be made quite different. The difference is governed by the determinants of the lattice filters. If the determinants are equal in all lattices, then the analysis and synthesis filter banks will also be equal. In that case the system is *unitary*. A condition for obtaining unitarity is that the lattice filters have coefficients related by $a_i = d_i = \cos \phi_i$ and $b_i = -c_i = \sin \phi_i$, $i = 0, 1, \ldots, N/2 - 1$. Then all determinants are equal to 1.

The rather strange relation between the prototype filter unit sample response and the lattice coefficients is forced upon the system to make its unit sample response consist of a sequentially modulated version of the prototype filter. An alternative is to modify the transform. For further details, see [3].

Generalization of Lattice PR Systems

It is very simple to generalize the result from the previous subsection to higher order systems. As we have derived earlier, the polyphase matrices of the analysis and synthesis systems consist of superlattices and the cosine transform. In the analysis-synthesis combination the cosine transforms are inverse systems, and so are the individual superlattices. For the specific example above the superlattices were replaced by simple lattices (Figure 3.47) $\mu_i(z)$. A superlattice may in fact be composed as a cascade of $\mu_i(z)$ in Equation 3.135, delay matrices $D_2(z^2)$ and standard 2×2 delay-free transforms

$\eta_{i,k}$ as:

$$\mathbf{M}_i(z) = \boldsymbol{\eta}_{i,q}\mathbf{D}_2(z^2)\boldsymbol{\eta}_{i,q-1}\mathbf{D}_2(z^2)\cdots\boldsymbol{\eta}_{i,1}\mathbf{D}_2(z^2)\boldsymbol{\mu}_i(z) \qquad (3.137)$$

The synthesis two-port is thus given by

$$\mathbf{M}_i^{-1}(z) = \boldsymbol{\mu}_i^{-1}(z)\mathbf{D}_2^{(i)}(z^2)\boldsymbol{\eta}_{i,1}^{-1}\cdots\mathbf{D}_2^{(i)}(z^2)\boldsymbol{\eta}_{i,q-1}^{-1}\mathbf{D}_2^{(i)}(z^2)\boldsymbol{\eta}_{i,q}^{-1} \qquad (3.138)$$

The inverses of the different components are given in Equations 3.87 and 3.89. Observe that $\boldsymbol{\mu}$ is a special case of $\boldsymbol{\eta}$. Analysis and synthesis super-lattice realizations based on primitive lattices are shown in Figure 3.49. In

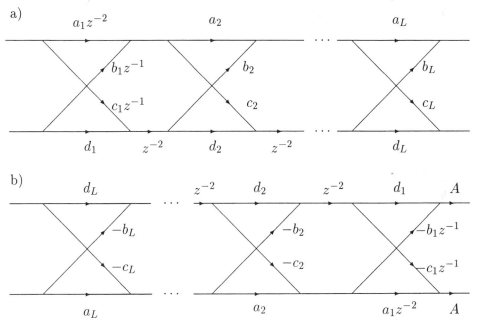

Figure 3.49: *Generalized two-port lattice structures for use in cosine-modulated filter banks. The indices i are left out for convenience. a) Analysis system. b) Synthesis system.*

Figure 3.49 b) the coefficients A_i are given by the product of the determinants of the analysis lattices as

$$A_i = \frac{1}{\det \boldsymbol{\mu}_i \prod_{k=1}^{q} \det \boldsymbol{\eta}_{i,k}}. \qquad (3.139)$$

The given circuits represent a structurally constrained PR system. Any coefficient set will render a PR system as long as no matrices are singular. A unitary system is obtained by selecting the lattice coefficients in Figure 3.49 as $a_i = d_i = \cos\phi_i$, and $b_i = -c_i = \sin\phi_i$. In this case the determinant product in Equation 3.139 $A_i = 1$, $i \in 0, 1, \ldots, N/2 - 1$.

Although the complexity of the cosine-modulated filter banks is very low - they may, in fact, have the lowest number of arithmetic operations possible for a filter bank with same number of channels and filter lengths - their performance in subband coding of images may not have the same virtue. As will become evident in Chapter 8, some of the important features in a well designed filter bank for image coding cannot be achieved by the presented model. The inherent structure is too constrained for the desired flexibility. From the design example above, we observe that none of the analysis highpass channels have very high attenuation at the origin (DC), something which is necessary to obtain good energy packing. Also the synthesis lowpass filter must have high attenuation around the frequencies $2\pi k/N$ for $k \in 1, 2, \ldots, N/2$ to suppress the dc imaging components. In these filters this can be obtained only by increasing the filter order at the expense of longer unit sample responses, which in turn may cause ringing problem for low bit rate coding.

The most interesting application seems to be high-rate to medium-rate coding where the complexity issue is crucial. In that case, however, a viable competitor may be the parallel allpass based filter banks from the previous section.

3.6.3   Extension of the Lattice Concept

The polyphase formulation is quite versatile and has proven a good aid in the search for PR systems. For the two-channel case we found rather general structures based on polyphase systems incorporating lattice structures cascaded with delay elements combined in a certain fashion.

Inspired by this, we may ask the question: do there exist similar N-channel systems with the same flexibility and built-in PR structure? A solution comes from the following simple idea: Compose an N-channel system from a cascade of delay-free $N \times N$ transforms and diagonal delay matrices. These are both easily invertible if we allow for an overall system delay resulting from using a causal pseudo-inverse of the delay matrix. More precisely, we describe the polyphase matrix as

$$\mathbf{P}(z) = \mathbf{C}_1 \mathbf{D}_{N,1}(z) \mathbf{C}_2 \mathbf{D}_{N,2}(z) \dots \mathbf{D}_{N,L-1}(z) \mathbf{C}_L, \qquad (3.140)$$

where all components $\{\mathbf{C}_i, i = 1, 2, \dots, L\}$ are z-independent coupling matrices and

$$\mathbf{D}_{N,i} = \text{diag}[z^{-k_i(1)}, z^{-k_i(2)}, \dots, z^{-k_i(N)}], \quad i = 1, 2, \dots, L - 1, \qquad (3.141)$$

are diagonal delay matrices where $k_i(j) \in Z_+$. In order to make useful systems the delay elements should not all be the same.

The analysis filter bank based on this idea including N-times critical downsampling is shown in Figure 3.50. The original delays (before using the noble identities) must be selected as integer multiples of N. If the delays are selected otherwise the subsamplers cannot be moved passed the delays.

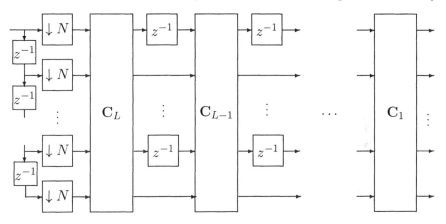

Figure 3.50: *Overall N-channel analysis filter bank based on cascades of delay-free matrices and diagonal delay matrices.*

In the figure we have made a special selection of the delay matrices where every second entry is equal to 1 and the others equal to z^{-1}. The synthesis structure includes the inverse of the polyphase matrix as

$$\mathbf{Q}(z) = \mathbf{P}^{-1}(z) = \mathbf{C}_L^{-1} \mathbf{D}_{N,L-1}^{(i)}(z) \cdots \mathbf{D}_{N,2}^{(i)}(z) \mathbf{C}_2^{-1} \mathbf{D}_{N,1}^{(i)}(z) \mathbf{C}_1^{-1}, \qquad (3.142)$$

where the causal pseudo-inverse of the delay matrices is given by

$$\mathbf{D}_{N,j}^{(i)}(z) = \mathbf{D}_{N,j}^{-1}(z) z^{\max(k(j))}. \qquad (3.143)$$

Here, $\max(k(j))$ represents the largest delay in the delay matrix $D_{N,j}(z)$.

The inverse filter bank can then be implemented as given in Figure 3.51.

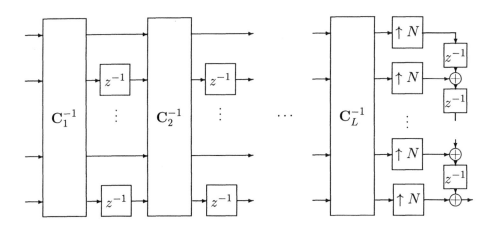

Figure 3.51: *Overall N-channel synthesis filter bank based on cascades of coefficient matrices and delay-matrices.*

Design Freedom

If we fully specify the analysis filter bank, there are no free parameters for the synthesis filter bank. Alternatively, we can use all the free parameters for the synthesis filter bank and thus leave no free parameters for the analysis filter bank. All intermediate cases are also possible. With some effort one might devise schemes where optimization is done both on the analysis and synthesis sides simultaneously, but then with reduced design flexibility for each of them.

The structure seems to be very flexible. First of all, the number of stages is at the liberty of the designer. Another choice is the generality of the coefficient matrices. Should they be sparse, thus reducing the computational burden for a given filter order? The delay matrices also leave a lot of flexibility. Again there is a tradeoff between filter order and generality. Here the cost by increasing the order is only more memory.

We will not in general try to indicate good tradeoffs in the design process. However, we exemplify a filter bank design for a 4-channel case. Figure 3.52 shows the frequency responses of the analysis filters where the coefficients in

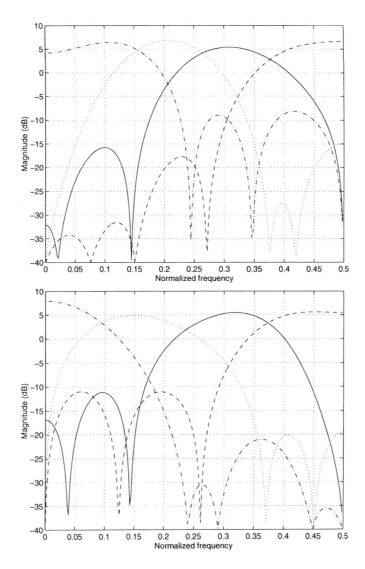

Figure 3.52: *Magnitude responses for the optimized 4-channel analysis filter bank (upper figure) and synthesis filter bank (lower figure).*

the transforms have been optimized for coding gain as explained in Chapter 5 (AR(1), $\rho = 0.95$). The filters are forced to have linear phase. The coding gain of this filter bank for the given process is 8.714 dB. The reader should compare the frequency characteristics and the gain in these filters with the ones given in Chapter 5.

3.7 Summary

We have demonstrated a series of analysis filter banks having minimum (Nyquist) representation and for which ideal inverses exist. Not only are all aliasing components that inevitably are generated during decimation cancelled out in the synthesis filter banks, all amplitude and phase distortions are eliminated as well. These are properties with great practical implications.

We organized the discussion into two main topics, the two-channel systems and the parallel systems. The two-channel filters can be used in tree-structured filter banks. Other natural distinctions among the filter banks are such as FIR versus IIR filters or a combination thereof, complex versus real signals, etc. One could also include near perfect reconstruction systems, which we have presented only in one special case.

The merits of the different filter banks are quite varied. Different applications and types of implementations may require entirely different algorithms, with tradeoffs like flexibility in design and minimization of cost. The best solution may be entirely different for say, a software implementation, a low-speed VLSI implementation, and a high-speed VLSI implementation. Two quite different applications are spatial filtering for still image coding, and temporal filtering for the additional dimension in 3-D subband video coding. We therefore need an arsenal of different filter banks.

There are definitely tradeoffs between design flexibility and complexity. The simplest filter banks are those with some inherent structure, either as tree structures, or the channels are interdependent because all filters are derived by modulation of a prototype filter. For some low bit rate applications the filter banks with the most structure (lowest degree of freedom for optimization), cannot avoid e. g. blocking artifacts for a given filter length whereas unstructured filter banks can give perfect results for the same filter lengths. In Chapters 7 and 8 we shall demonstrate some of these effects. The last word is definitely not said about these matters. Although not treated extensively in this book, we believe at the time of writing that *adaptive filter banks* may see an interesting future for further subband coder optimization.

Chapter 4

Bit Efficient Quantization

Signal decomposition does not provide rate reduction by itself. To obtain low rate, *quantization* of the signal, or parameters derived from the signal, is necessary. The detailed way of representing a complete signal using a finite number of levels is the task of quantization, and the main topic of this chapter.

The organization of the chapter is as follows: The next two sections discuss the fundamentals of quantization. Both individual quantization of the signal components (*scalar quantization*) as well as *vector quantization* (VQ) will be covered. In Section 4.3 we review the bit and level allocation problem, where we try to distribute the bits according to the importance of the different parameters. Following this derivation we turn to the *entropy coding* method and its use for representing decomposed sources. Comparisons between the bit allocation, and the entropy coding methods will be given.

4.1 Scalar Quantization

The simplest form of digital representation is obtained by scalar quantization. Scalar quantization can be subdivided in several methods ranging from uniform quantization to quantization optimized to the probability density function of the input signal.

A *scalar* quantizer $Q[\cdot]$ can in general be thought of as a mapping of $x \in \mathbf{R}$ to a finite set $Y = \{y_0, \ldots, y_{L-1} | y_i \in \mathbf{R}\}$:

$$Q : \mathbf{R} \to Y. \tag{4.1}$$

An example of a quantizer mapping is shown in Figure 4.1.

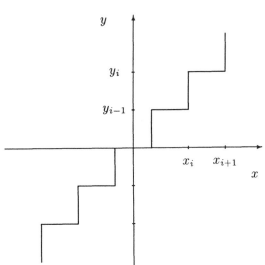

Figure 4.1: *Uniform quantizer mapping.*

The mapping Q[·] is always defined on a *partition* of **R** into nonoverlapping *intervals* $\{R_i\}$ according to the rule

$$x \in R_i \quad \Longrightarrow \quad Q[x] = y_i, \quad i = 0, 1, \ldots, L - 1, \tag{4.2}$$

where $R_i = (x_i, x_{i+1})$ and satisfies

$$\bigcup_{i=0}^{L-1} R_i = \mathbf{R} \quad \text{and} \quad R_i \cap R_j = \emptyset \quad \text{for} \quad i \neq j. \tag{4.3}$$

Here, $(\cdot, \cdot)$ denotes an open, half open or a closed interval. The values on the vertical axis of the input/output mapping of Figure 4.1 correspond to elements $\{y_i, \ i = 0, 1, \ldots, L - 1\}$ of Y and are referred to as *representation levels*. The associated values on the horizontal axis, denoted $\{x_i\}$, are referred to as *decision levels*.

In practice we have to split the operations in an *encoder* which performs the mapping

$$C[x] = i, \quad i \in 0, 1, \ldots, L - 1, \tag{4.4}$$

where the index i points to the quantization interval, and a *decoder* with the mapping

$$D[i] = y_i, \quad i = 0, 1, \ldots, L - 1, \tag{4.5}$$

where the index i identifies the representation value.

For the mapping of Figure 4.1, the representation levels are the midpoints between the decision levels, and the decision interval length does not vary within the active quantizer region. Such a quantizer is referred to as a *uniform* quantizer. In the coding experiments of this book, we will frequently make use of uniform quantization, but with a *dead-zone* around the zero level. Figure 4.2 depicts a uniform quantizer where the input samples with absolute values less than t are set to zero.

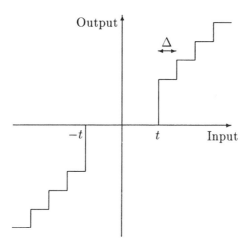

Figure 4.2: *Uniform quantization combined with a "dead-zone".*

An example of a more general, *nonuniform* scalar quantizer mapping is depicted Figure 4.3.

For signals distributed according to a uniform probability density function (pdf), the optimal (in the mean square sense) *scalar* quantizer is a uniform quantizer. If the signal pdf is not uniform, we can use a pdf-optimized or an entropy constrained quantizer to obtain better performance. These schemes are described in the following two subsections.

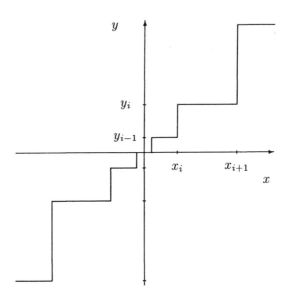

Figure 4.3: *Nonuniform quantizer mapping.*

4.1.1   The Lloyd-Max Quantizer

The Lloyd-Max quantizer is a scalar quantizer where the 1st order pdf of the signal to be quantized is exploited to increase the performance. It is therefore often referred to as a *pdf-optimized quantizer*. Each signal sample is quantized using the same number of bits. The optimization is done by minimizing the total distortion of a quantizer with a given number L of representation levels. For an input signal x with pdf $p_x(x)$, using the notation introduced in the previous section, the mean square distortion can be calculated from

$$\sigma_q^2 = \sum_{i=0}^{L-1} \int_{x_i}^{x_{i+1}} (x - y_i)^2 p_x(x) dx. \tag{4.6}$$

Minimization of σ_q^2 was done by Max [54] and Lloyd [55], and leads to the following expressions for the decision and representation levels:

$$x_{k,opt} = \frac{1}{2}(y_{k,opt} + y_{k-1,opt}), \qquad k = 1, 2, \ldots, L - 1 \tag{4.7}$$

$$x_{0,opt} = -\infty \tag{4.8}$$

$$x_{L,opt} = \infty \tag{4.9}$$

$$y_{k,opt} = \frac{\int_{x_{k,opt}}^{x_{k+1,opt}} x p_x(x) dx}{\int_{x_{k,opt}}^{x_{k+1,opt}} p_x(x) dx}, \qquad k = 0, 1, \ldots, L-1. \tag{4.10}$$

The solution does not give an explicit formula for the quantization mapping. Equation 4.7 simply indicates that the decision levels should be the midpoint between neighboring representation levels, while Equation 4.10 states that the optimal representation levels are the *centroids* of the pdf in the appropriate interval.

The equations can be used in an iterative manner to obtain the optimal representation and decision levels [10].

High rate approximation

At high bit rates it is possible to derive approximate formulas for the pdf-optimized quantizer based on so-called *compander characteristics* [10]. For this case the quantization noise is accurately described by

$$\sigma_q^2 = \epsilon_x \frac{\sigma_x^2}{L_x^2}, \tag{4.11}$$

where ϵ_x can be calculated from

$$\epsilon_x = \frac{1}{12} \left\{ \int_{-\infty}^{\infty} [p_x(x)]^{1/3} dx \right\}^3, \tag{4.12}$$

where $p_x(x)$ is the pdf of a *unit variance* stochastic variable x. In Table 4.1 numerical values for ϵ_x is tabulated for some well-known distributions.

In the derivation of optimal bit allocation for sources with mixture statistics in Section 4.3.1, these expressions will serve as quantizer models.

It is important to observe that pdf-optimized quantizers produce identical contributions to the error variance from each quantizer interval. However, the quantizer intervals are not equiprobable (except for the uniform pdf case).

When minimizing the total distortion for a fixed number of possible representation levels, we have tacitly assumed that every signal sample is coded using the same number of bits: $\log_2 L$ bits/sample. Because of the different occurence probability of each quantizer interval, a further reduction in the total number of bits used may be achieved by assigning few bits to highly probable representation levels and many bits to less probable representation levels. This invariably leads to variable bit rate. When this possibility exists, it turns out that the Lloyd-Max solution is no longer optimal. Therefore a new optimization is needed, leading to the *entropy constrained quantizer*.

4.1.2   Entropy Constrained Quantization

Instead of minimizing the total distortion for a fixed number of representation levels, we minimize, for a fixed distortion, the 1st order entropy of the discrete source associated with the quantizer output. The rationale for this procedure is the Shannon noiseless source coding theorem [5], which states that the 1st order entropy is a lower bound for the achievable average bit rate when the signal samples are quantized and converted into bit patterns in an independent manner.

At high bit rates, it can be shown [10] that the optimum is reached when using a uniform quantizer with an infinite number of levels. At low bit rates, it was shown in [56], that *uniform midtread threshold quantizers* perform close to optimum for some common sources. A uniform threshold quantizer uses uniformly spaced decision levels, while the representation levels are the centroids of each quantization interval.

Let us us calculate the entropy of a uniformly quantized signal. This will be used as a lower bound for the practically obtainable bit rate.

Recall the expression for the 1st order entropy measured in bits/sample:

$$H(X) = - \sum_{i=0}^{L-1} P_i \log_2(P_i), \tag{4.13}$$

where P_i is the probability of representation level y_i. If we associate with y_i a codeword or bit pattern of length l_i bits, the average codeword length is given by

$$\bar{l} = \sum_{i=0}^{L-1} P_i l_i. \tag{4.14}$$

Comparing the two above equations, we see that the entropy bound is reached if we select the codeword lengths to $l_i = - \log_2(P_i)$.

As we are going to entropy code quantized versions of analog signals, it is convenient to express the entropy by the pdf of the analog signal plus parameters of the quantizer. We consider only uniform quantizers for which the quantization intervals are all equal, $\Delta = x_{i+1} - x_i$, $i = 0, 1, \ldots, L - 1$. We then make an approximation, which is accurate when the step size is so small that the probability density function is almost constant over that interval,

$$P_i \approx p_x(y_i)\Delta. \tag{4.15}$$

$p_x(x)$ is the probability density function of the analog source. The first order entropy can then be approximated by

$$
\begin{aligned}
H(X) &\approx -\sum_{i=0}^{L-1} p_x(y_i)\Delta \log_2(p_x(y_i)\Delta) \\
&= -\sum_{i=0}^{L-1} p_x(y_i) \log_2(p_x(y_i))\Delta - \log_2(\Delta) \\
&\approx -\int_{-\infty}^{\infty} p_x(x) \log_2(p_x(x))dx - \log_2(\Delta) \\
&= h(X) - \log_2(\Delta),
\end{aligned}
\tag{4.16}
$$

where $h(X)$ is called the *differential entropy*.

The differential entropy can be calculated explicitly for certain *memoryless sources* and has the form

$$h(X) = \frac{1}{2}\log_2(\eta_x \sigma_x^2), \tag{4.17}$$

where η_x depends on the pdf of the source and σ_x^2 is the source variance. Numerical values are presented for certain common sources in Table 4.1 [10]. For the given sources η_x is independent of the source variance. For other sources this independence does not necessarily exist. We still maintain this form and define in general η_x[1] as the solution of Equation 4.17:

$$\eta_x = \frac{1}{\sigma_x^2} 2^{2h(X)}. \tag{4.18}$$

Using Equations 4.16 and 4.17 the quantized signal entropy can be expressed as

$$H(X) = \frac{1}{2}\log_2[\eta_x(\frac{\sigma_x}{\Delta})^2]. \tag{4.19}$$

---

[1] η_x is related to the *entropy power* Q_x through the relation $\eta_x = 2\pi e \frac{Q_x}{\sigma_x^2}$.

The 1st order entropy only depends on η_x and on the ratio between the standard deviation and the quantization interval.

Now assume that the signal samples to be quantized are independent and identically distributed according to the pdf $p_x(x)$. For high bit rates, where the assumption of flat pdf within each quantization interval Δ is justified, the mean square distortion can be estimated as

$$\sigma_q^2 = \frac{\Delta^2}{12},$$ (4.20)

and the minimal 1st order entropy $H(X)$ can be expressed in terms of the *rate distortion lower bound* $R(\sigma_q^2)$ [10]:

$$H(X) = R(\sigma_q^2) + 0.255.$$ (4.21)

Equation 4.21 tells us that irrespective of signal pdf, the price paid for confining ourselves to scalar quantization is only 0.255 bits/sample extra for *memoryless sources*. It should be emphasized that this applies to extremely idealized conditions: In a subband coder, the subband signals have some inherent dependencies (2nd or higher order). This leads to a rate distortion lower bound that is smaller than for independent subband samples. Also, we repeat that the 1st order entropy is a *lower bound* for the achievable bit rate in a scalar quantizer. Practical coders like *Huffman coders* [57] or *arithmetic coders* typically produce an average bit rate that is somewhat higher than that given by the entropy.

Pdf-Optimized versus Entropy-Constrained Scalar Quantizers

In Table 4.1 expressions and numerical values for η_x are given for some well-known sources. In the same table numerical values for the quantizer performance factor ϵ_x (from Equation 4.12) are given for the same sources [10].

The difference in high bit rate performance between pdf-optimized scalar quantizers and entropy-constrained quantizers can be found in the following way: Find the size of the quantization interval used in a uniform quantizer which will produce the same noise as the pdf-optimized quantizer by equating Equations 4.11 and 4.20

$$\epsilon_x \frac{\sigma_x^2}{L_x^2} = \frac{\Delta^2}{12}.$$ (4.22)

pdf	η_x	ϵ_x	$\frac{\eta_x}{12\epsilon_x}$	$b - H(X)$
Uniform	12	1	1	0
Gaussian	$2\pi e = 17.07$	$\frac{\pi}{2}\sqrt{3} = 2.72$	0.522	0.47
Laplacian	$2e^2 = 14.78$	$\frac{9}{2} = 4.50$	0.274	0.93
Gamma	$\frac{4}{3}\pi e^{1-C} = 6.39$	$4\sqrt{\frac{3}{\pi}}\Gamma^3(\frac{5}{6}) = 5.62$	0.095	1.70

Table 4.1: *Values of the constants η_x and ϵ_x for some common sources. $C = 0.5772$ is Euler's constant. Also the ratio $\eta_x/(12\epsilon_x)$ is given together with the corresponding rate advantage of the entropy-constrained quantizer compared to the pdf-optimized quantizer valid for high rates.*

Solving this equation with respect to Δ^2 and inserting the result into Equation 4.19, we find the relation between the entropy, which is an estimate of the rate in the entropy-constrained quantizer $H(X)$, and the bits per sample $b = \log_2(L_x)$ used for the pdf-optimized quantizer:

$$H(X) = \frac{1}{2}\log_2(\eta_x \frac{\sigma_x^2}{\Delta^2}) = \frac{1}{2}\log_2(\frac{\eta_x}{\epsilon_x}\frac{L_x^2}{12}) = b + \frac{1}{2}\log_2(\frac{\eta_x}{12\epsilon_x}). \qquad (4.23)$$

Values of this ratio is given in Table 4.1. The rate advantage of the entropy-constrained quantizer over the pdf-optimized quantizer is given in the last column in the table.

Although this result is somewhat optimistic with respect to the rate of the entropy-constrained quantizer, the peaky distributions clearly favor this representation method.

4.2   Vector Quantization

Until now we have discussed *scalar* quantization techniques where the signal samples have been quantized independently. Simultaneous quantization of several samples is referred to as *vector quantization* (VQ) [7]. The concept of VQ is a generalization of scalar quantization: A vector quantizer Q[·] is a mapping of $\mathbf{x} \in \mathbf{R}^N$ to a finite set $\mathbf{Y} = \{\mathbf{y}_0, \ldots, \mathbf{y}_{L-1} | \mathbf{y}_i \in \mathbf{R}^N\}$ of representation vectors:

$$Q : \mathbf{R}^N \to \mathbf{Y}. \qquad (4.24)$$

Similar to scalar quantization, the mapping $Q[\cdot]$ can be defined on a *partition* of $\mathbf{R}^N$ into nonoverlapping, N-dimensional *cells* $\{C_i\}$ according to the rule

$$\mathbf{x} \in C_i \quad \Longrightarrow \quad Q[\mathbf{x}] = \mathbf{y}_i, \quad i = 0, 1, \ldots, L-1 \qquad (4.25)$$

where $\{C_i\}$ satisfies

$$\bigcup_{i=0}^{L-1} C_i = \mathbf{R}^N \quad \text{and} \quad C_i \cap C_j = \emptyset \quad \text{for} \quad i \neq j. \qquad (4.26)$$

It is seen that the vectors $\{\mathbf{y}_i\}$ correspond to the representation levels in a scalar quantizer. In a VQ setting the collection of representation levels is referred to as the *codebook*. The cells C_i, also called *Voronoi regions*, or *Dirichlet partitions*, correspond to the decision levels, and can be thought of as solid polygons in the N-dimensional space $\mathbf{R}^N$. In the scalar case it is trivial to test if a signal sample belongs to a given interval. In VQ an indirect approach is to utilize a *fidelity criterion* or *distortion measure* $d(\cdot, \cdot)$:

$$Q[\mathbf{x}] = \mathbf{y}_i \iff d(\mathbf{x}, \mathbf{y}_i) \leq d(\mathbf{x}, \mathbf{y}_j), \qquad j = 0, 1, \ldots, L-1. \qquad (4.27)$$

When the best match, $\mathbf{y}_i$, has been found, the *index i* identifies that vector and is therefore coded as an efficient representation for the vector. The receiver can then reconstruct the vector $\mathbf{y}_i$ by looking up the representation vector in cell number i in a copy of the codebook. Thus, the bit rate in bits per sample in this scheme is $\frac{1}{N} \log_2 L$, when using straight forward bit representation for i. The principle of vector quantization is shown in Figure 4.4. Note that the different VQ indices are not equiprobable as was also the case for pdf-optimized scalar quantizers. A further bit reduction can thus be obtained by entropy coding of the indices.

In the previous section we stated that scalar entropy coding was suboptimal, even for a source producing independent samples (Equation 4.21). At first, this looks rather strange since we seem to have exploited all redundancy in the signal: The signal samples to be quantized were assumed to be totally independent, and the quantizer was adapted to the pdf of the signal. The reason for the suboptimal performance of the entropy constrained quantizer is a phenomenon called *sphere packing*. To illustrate the problem, let us assume that we want to quantize a signal $x(n)$ with *uniformly distributed* and *independent* samples. The optimal scalar quantizer is a uniform

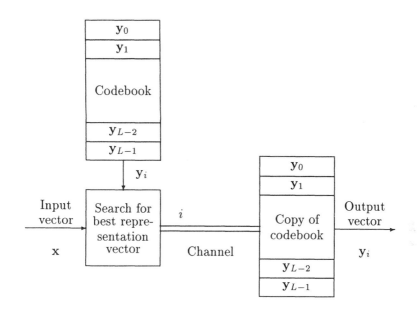

Figure 4.4: *Vector quantization procedure.*

quantizer (in this case all output values have the same probability, thus no entropy coding is necessary). Therefore each input sample is mapped according to which *interval* it belongs to. If we split the input signal into a collection of two-dimensional vectors according to $\mathbf{w}_n = (x(2n), x(2n+1))^T$, the scalar quantization procedure can be viewed as a partition of the plane into *square boxes*, or *cells*: Each $\mathbf{w}_n$ is quantized according to which cell it belongs. Figure 4.5 a) illustrates the resulting partitioning of the plane. The dots shown are the representation vectors and the cell walls are the decision region boundaries. The *hexagon cell partition* of Figure 4.5 b) exhibits lower mean quantizer distortion for a fixed number of cells than that of the square or rectangular partition, and can be shown [58] to be optimal for $\mathbf{R}^2$ if the signal is uniformly distributed.

Similarly, for higher dimensions the *sphere packing capabilities* of VQ will give an increased gain. Theoretical bounds for the case of independently, uniformly distributed N-dimensional vectors are given in [59]. Thus, if we restrict ourselves to scalar quantization, the associated N-dimensional VQ partition of $\mathbf{R}^N$ will always consist of *rectangular* N-dimensional cells - and suboptimality (Equation 4.21) results.

a)   b)

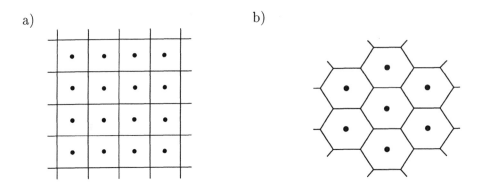

Figure 4.5: *2-dimensional sampling grids: a) Scalar (left), b) Optimal (right).*

Now assume that we want to design a codebook for a VQ scheme where the N-dimensional pdf is known. Luckily, finding the geometry of the optimal N-dimensional cells is not necessary. Instead one should note that the shape of each cell (which may vary within the partition) is uniquely determined by the distortion measure $d(\cdot, \cdot)$ and the representation vectors in the codebook. Therefore, given a distortion measure, we have to find the optimal representation vectors to be used in the codebook. In practice, the multidimensional pdf is unknown. Instead a *training set* of vectors is used. This set consists of a large number of vectors that are representative for the signal source. A suboptimal codebook can then be designed using an iterative algorithm, for example the *K-means* or *LBG* algorithm [60].

In addition to obtaining good sphere packing in $\mathbf{R}^N$, a VQ scheme will also exploit both *linear* (i.e. correlation) and *nonlinear* or *higher order statistical dependency* that may be inherent in a multidimensional pdf. The higher order dependency can be thought of as "a preference for certain vectors". As an example, consider a two-dimensional pdf where the support lies on a square frame centered about the origin, see Figure 4.6. Assuming the pdf to be a constant on the frame, the reader may assure herself/himself that the two variables within the vector indeed are uncorrelated. A scalar quantizer will waste bits because many of the representation vectors of the associated rectangular partition of $\mathbf{R}^2$ will lie *within* the frame where the pdf is zero. These representation vectors are never used. In a good VQ codebook all legal representation vectors would lie *on* the frame.

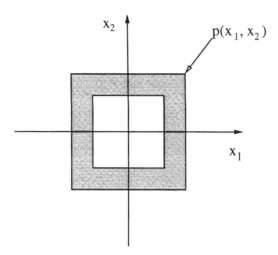

Figure 4.6: *2-D pdf example where the variables are uncorrelated, but depen-dent.*

Finally, we would like to add that even though a scalar quantizer is in-capable of exploiting correlation between signal samples, decorrelation tech-niques such as Karhunen-Loève transformation, DPCM, or indeed subband filtering can be applied *prior* to quantization. Thus, for subband coding, the extra gain offered by VQ is due to exploitation of higher order dependencies and sphere packing. For low rates these two gains are small [61].

4.3 Scalar Quantization and Mixture Distributions

Although VQ is optimal for coding vectors, its complexity becomes pro-hibitive as the vector length increases, or more accurately, as the codebook size increases. A much simpler method, which can also easily adapt to non-stationarities in the signal statistics, uses varying number of bits for different signal components (e.g. after signal decomposition) according to some im-portance measure that can be derived from the signal. It will be shown in Chapter 5 that signal decomposition can remove correlation and facilitate the use of scalar quantization or simple vector quantizers.

The following represents a naïve conception of adaptive use of bit re-sources: Any signal consists of small and large amplitudes. In standard scalar quantization every sample is represented by the same word length.

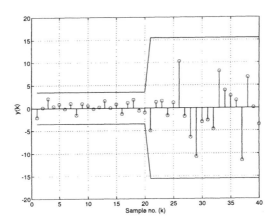

Figure 4.7: *Signal arranged in two size groups. The signal is Gaussian with variance 1 for the first 20 samples and variance 16 for the last 20 samples.*

Provided the data samples could be ranked by size, and grouped into certain size classes, each class could use different quantizers with an appropriate number of bits to accommodate for the dynamic range and the necessary resolution in each class. An example of a signal with two distinct size groups is shown in Figure 4.7. If we use a quantization interval $\Delta = 1$, then 3 and 5 bits, respectively, suffice to represent the two groups if the saturation levels are set to ± 3.5 and ± 15.5, for a midrise quantizer, as indicated in the figure.

In order to make the organization of signal amplitudes useful in a coding context, the rearrangement done in the encoder must be known to the receiver for making a correct interpretation of the bit stream. Two obvious solutions exist:

- A rearrangement table must be supplied as side information to the receiver.

- The rearrangement must be done systematically in a way known to the receiver.

The first solution requires an excessive amount of bits if done sample by sample. The second solution is relevant for stationary signals after decomposition. The obtained components would then typically have different sizes depending on the shape of the input signal spectral density. If arranged systematically, the receiver would know the bit distribution among the components.

In the following we consider methods for quantizing the collection of samples with different statistics. Let us assume that we generate a signal y by picking samples from N memoryless and independent sources with pdfs $\{p_{y_n}(y_n), \ n = 0, 1, \ldots, N - 1\}$. We assume that all components are *zero mean*[2]. We can now characterize y by the *mixture distribution*[62]:

$$p_y(y) = \sum_{n=0}^{N-1} P_n p_{y_n}(y), \qquad (4.28)$$

where P_n is the fraction of samples taken from the source y_n. If we furthermore form a sequence by picking samples in a random order from the different sources, y becomes stationary. If picking samples in a systematic order, we generate a *cyclostationary* sequence.

The variance of this new pdf is given by

$$\sigma_y^2 = \int_{-\infty}^{\infty} y^2 \sum_{n=0}^{N-1} P_n p_{y_n}(y) dy = \sum_{n=0}^{N-1} P_n \sigma_{y_n}^2. \qquad (4.29)$$

An example of two Gaussian pdfs with variances equal to 1 and 16 are shown in Figure 4.8 together with the mixture distribution composed of an equal mixture of both pdfs ($P_0 = P_1 = 0.5$). The variance of the mixture distribution is $\sigma_y^2 = (1 + 16)/2 = 8.5$.

Direct coding of the mixture distribution is very different from separate coding of each component, and in fact, much more difficult. This is due to the shape of the pdf of the mixture distribution, as we will experience in the sequel.

Let us collect one sample from each of the N channels after subband filtering into a vector $\mathbf{y}(l) = [y_0(l), y_1(l), \ldots, y_{N-1}(l)]$. If viewing the vector components as sample values of a mixture source y, the pdf of this source is given by

$$p_y(y) = \frac{1}{N} \sum_{n=0}^{N-1} p_{y_n}(y). \qquad (4.30)$$

---

[2]For subband samples this assumption is justified for all bands, due to bandpass filtering, except the lowpass band.

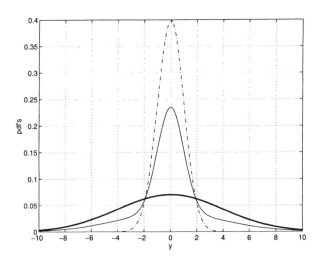

Figure 4.8: *Pdfs for two Gaussian signals with variances 1 (dotted dashed line) and 16 (thick solid line) plus a mixture distribution (thin solid line) composed of an equal number of samples from both distributions.*

In this case there is an equal number of samples from each source represented in the new source and consequently $P_i = 1/N$.

The individual coding of the vector components is facilitated by the way the samples are generated in a subband coder. Each component will be generated through filtering in one bandpass filter. We will consider this problem in great detail in the next sections.

4.3.1　Optimal Bit Allocation

In this section we look at the simultaneous *scalar* quantization of a vector $\mathbf{y} = [y_0, y_1, \ldots, y_{N-1}]$ using a total of B bits while allowing for different no. of bits for each component. The optimization criterion we will employ is the minimization of the average distortion.

Denoting the quantized signal components $Q_n(y_n)$, the average distortion per component can be written as

$$\sigma_d^2 = \frac{1}{N} \sum_{n=0}^{N-1} E[(y_n - Q_n(y_n))^2] = \frac{1}{N} \sum_{n=0}^{N-1} \sigma_{q_n}^2, \qquad (4.31)$$

where $E[\cdot]$ is the expectation operator and $\sigma_{q_n}^2$ is the quantization noise variance in component n. The bit constraint is given by

$$B = \sum_{n=0}^{N-1} b_n, \tag{4.32}$$

where b_n is the number of bits used to quantize component no. n. Denoting the number of representation levels used for quantizing vector element no. n by $L_n = 2^{b_n}$, we can rewrite Equation 4.32 as

$$B = \sum_{n=0}^{N-1} \log_2(L_n) = \log_2 \left(\prod_{n=0}^{N-1} L_n \right). \tag{4.33}$$

Then

$$2^B = \prod_{n=0}^{N-1} L_n = L_g^N, \tag{4.34}$$

where we have introduced the geometric mean, L_g, of the number of levels used for the different components. This equation represents a *level constraint*: The product of the number of levels to be used for the vector is fixed and equal to L_g^N.

In order to obtain the proper bit/level assignment, we assume that the quantizer performance can be modeled by Equation 4.11 for each component, that is

$$\sigma_{q_n}^2 = \epsilon_{y_n} \frac{\sigma_{y_n}^2}{L_n^2}. \tag{4.35}$$

We now minimize the average distortion given in Equation 4.31 using the proper distortions in Equation 4.35 subject to the level constraint in Equation 4.34 with respect to the number of levels in quantizer i. Using the Lagrange multiplier method [63], we require:

$$\frac{\partial}{\partial L_i} \left\{ \frac{1}{N} \sum_{n=0}^{N-1} \epsilon_{y_n} \frac{\sigma_{y_n}^2}{L_n^2} + \lambda \prod_{n=0}^{N-1} L_n \right\} = 0. \tag{4.36}$$

After differentiation and some rearrangement, we obtain

$$\sigma_{q_i}^2 = \epsilon_{y_i} \frac{\sigma_{y_i}^2}{L_i^2} = \frac{\lambda N L_g^N}{2}. \tag{4.37}$$

As $\lambda N L_g^N / 2$ is a constant (for minimum noise), *optimal level allocation implies equal noise in each component.*

Because all noise components are equal, we can write

$$\sigma_{q_i}^2 = \left(\prod_{n=0}^{N-1} \sigma_{q_n}^2\right)^{\frac{1}{N}} = \frac{\left(\prod_{n=0}^{N-1} \epsilon_{y_n}\right)^{\frac{1}{N}} \left(\prod_{n=0}^{N-1} \sigma_{y_n}^2\right)^{\frac{1}{N}}}{\left(\prod_{n=0}^{N-1} L_n^2\right)^{\frac{1}{N}}} = \epsilon_{y_g} \frac{\sigma_{y_g}^2}{L_g^2}, \qquad (4.38)$$

where we have introduced the following geometric means:

$$\sigma_{y_g}^2 = \left(\prod_{n=0}^{N-1} \sigma_{y_n}^2\right)^{\frac{1}{N}} \text{ and } \epsilon_{y_g} = \left(\prod_{n=0}^{N-1} \epsilon_{y_n}\right)^{\frac{1}{N}}. \qquad (4.39)$$

From Equations 4.37 and 4.38 we get

$$\epsilon_{y_g} \frac{\sigma_{y_g}^2}{L_g^2} = \epsilon_{y_i} \frac{\sigma_{y_i}^2}{L_i^2}, \qquad i = 0, \ldots, N-1. \qquad (4.40)$$

Solving with respect to the number of levels in component i, we find

$$L_i = L_g \sqrt{\frac{\epsilon_{y_i}}{\epsilon_{y_g}} \frac{\sigma_{y_i}^2}{\sigma_{y_g}^2}}, \qquad (4.41)$$

which is the optimal number of levels for component i.

We have derived the optimal level allocation as a function of the component variances. This gives more flexibility than bit allocation. However, many practical systems require that we stick to bits. The equivalent bit allocation formula can be easily found as

$$b_i = \log_2(L_i) = \frac{B}{N} + \frac{1}{2} \log_2\left(\frac{\epsilon_{y_i}}{\epsilon_{y_g}} \frac{\sigma_{y_i}^2}{\sigma_{y_g}^2}\right), \qquad i = 0, 1, \ldots, N-1. \qquad (4.42)$$

If all the components are identically distributed, then $\epsilon_{y_i}/\epsilon_{y_g} = 1$.

Finally, the minimum average distortion per component can be written as

$$\sigma_d^2 = \frac{1}{N} \sum_{n=0}^{N-1} \sigma_{q_n}^2 = \epsilon_{y_g} \frac{\sigma_{y_g}^2}{L_g^2}. \qquad (4.43)$$

Let us now compare this distortion to the distortion when quantizing all components with the same scalar quantizer. That is, we quantize y as one source that can be be characterized by the mixture distribution in Equation 4.28. Such quantization is usually called *pulse code modulation (PCM)*. The bit rate is equal to $b = B/N$ bits for all components, or equivalently L_g levels per component. The average signal power per component is given by σ_y^2. The PCM distortion per sample is accordingly

$$\sigma_{PCM}^2 = \epsilon_y \frac{\sigma_y^2}{L_g^2} = \frac{\epsilon_y}{L_g^2} \frac{1}{N} \sum_{n=0}^{N-1} \sigma_{y_n}^2, \tag{4.44}$$

provided a pdf-optimized quantizer for the mixture source is applied.

The *coding gain* using optimal bit-allocation (BA) is defined as the distortion advantage of the component-wise quantization over PCM quantization at the same rate, and can be calculated from the inverse ratios between the distortions as

$$G_{BA} = \frac{\sigma_{PCM}^2}{\sigma_d^2} = \frac{\epsilon_y}{\epsilon_{y_g}} \frac{\frac{1}{N}\sum_{n=0}^{N-1}\sigma_{y_n}^2}{\left[\prod_{n=0}^{N-1}\sigma_{y_n}^2\right]^{\frac{1}{N}}}. \tag{4.45}$$

The gain is seen to be equal to the ratio between the quantizer performance factor of the mixture source, ϵ_y, and the geometric mean, ϵ_{y_g}, of the components' performance factors, times the ratio between the mixture source variance, σ_y^2, which is also the arithmetic mean of the component variances, and their geometric mean, $\sigma_{y_g}^2$. This gain may not be very relevant. It would be an extremely bad idea to do signal decomposition and then consider the obtained components as a single (mixture) source that could be quantized by scalar methods. In fact, to evaluate the performance of a subband coder we use as a reference PCM quantization of the *input* signal. We define the subband coding gain in Chapter 5.

Example : Two-component source

We illustrate the two contributions to the gain G_{BA} of Equation 4.45 for a two-component example. Both components are Gaussian, one with variance $\sigma_0^2 = 1$, and the other with the variance σ_1^2, which is our free variable in the experiment. Figure 4.9 shows the two contributions and the total gain as a function of σ_1^2 in the range 1 to 25.

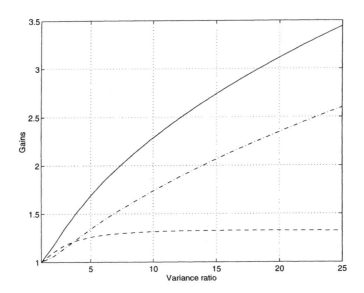

Figure 4.9: *Gains using optimal bit allocation in a two-component source. Dashed line: gain from quantizer performance factors. Dotted-dashed line: Gain from variance relations ("classical gain"). Solid line: Total gain.*

As can be observed from the figure, the contribution from the quantizer performance factors varies significantly in the lower range of the variance ratio. However, it decays again if extending the range as shown in Figure 4.10.

We refrain from presenting examples with several components due to the illustration problem. The tendency is clear from the example: Both factors in the gain formula do vary as a function of the variances.

□

Dynamic Bit Allocation

In the above theory, it has tacitly been assumed that the signals are *stationary*. For *nonstationary* sources the signal statistics would vary from one vector **y** to the next. It then becomes necessary to apply *dynamic bit allocation*. That is, the previous theory has to be applied to each block of data. However, the decoder must receive information about the local bit allocation. This can be done either by transmitting a *bit allocation table*,

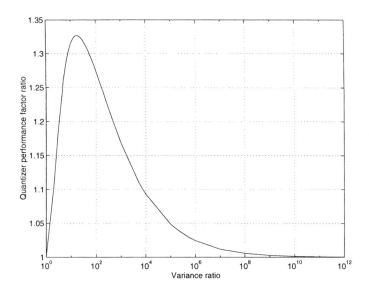

Figure 4.10: *Quantizer performance factor gain using optimal quantizers in a two-component source.*

or quantized versions of the variances that the bit allocation was derived from. In both cases the transmission of side information is required. For real images where the statistics vary rapidly, the cost of transmitting the side information may become costly, especially for low rate coders.

Practical Problems with the Bit Allocation Algorithm

A formula like Equation 4.42 may produce negative and noninteger bit numbers. Negative number of bits cannot be realized, and noninteger bit assignment would require block quantization.

A possible solution to this problem is to use some sort of a *greedy* algorithm [7] for which the bits are allocated iteratively according to some competition rule between the signal components.

The general algorithm can be explained as follows: Initially all components have zero bits ($b_n = 0, \quad n = 0, 1, \cdots, N - 1$), and each component is associated with some competitive advantage factor W_n. Then the algorithm goes as follows:

1. The component with maximum advantage factor, say component k, is assigned one bit: $b_k = b_k + 1$.

2. Reduce the advantage factor for component k according to some given rule, $W_k = W_k(b_k)$.

3. Compute the total number of bits allocated so far: $B_{acc} = \sum_{n=0}^{N-1} b_n$. While $B_{acc} < B$, where B is the total number of bits, go to 1. Otherwise stop.

We see that this kind of algorithm obviously guarantees both nonnegative and integer bit allocation to each component. It is also flexible as to the choice of the function W.

One such algorithm uses the standard deviation of each component as the initial advantage factor: $W_k(b_k) = \sigma_{y_k}/r^{b_k}$ [64]. This can be shown to correspond to Equation 4.42 when allocating the first few bits when selecting $r = 2$. The deviation occurs when we start running out of bits [65].

In the greedy algorithm it is possible to take into account the true distortion functions instead of the high-rate approximation given in Equation 4.11, which is inaccurate at low bit rates, as well as different pdfs in each band.

The drawback of the greedy algorithm is its computational complexity. Although each calculation is simple, there is one calculation per bit. At high bit rates, the number of calculations may become prohibitive. It is possible to speed up this process at the cost of introducing more complex calculations. One such scheme is to allocate a large percentage of the bits using Equation 4.42 and finally allocate the remaining bits with the greedy algorithm as described above.

4.3.2   Entropy Coder Allocation

In this section we return to entropy coding methods, and investigate how they can be applied to sources with mixture statistics. In Section 4.1.2 entropy coders were proven superior to pdf-optimized quantizers with fixed rate per symbol. How does this result carry over to this more complex case?

The rate of a source is lower bounded by its entropy. For discrete sources the entropy per symbol is finite and can be expressed by Equation 4.13, while the entropy of uniformly quantized continuous-amplitude-sources is fairly accurately expressed by Equation 4.19 for high rates.

In a mixture source y, obtained by picking an equal number of samples from N different sources with different statistics, it may be worth while to examine the possible advantage of allocating different *entropy coders* to the different sources based on their statistics, in much the same way as allocating different quantizers to the components in the bit allocation case. To evaluate the merits of such a scheme, we proceed to calculate the entropy of the mixture source y as well as the average rate of the components of this source. The results will also enable us to make comparisons between the bit allocation method and the two different entropy coding strategies.

Rate of a Mixture Source

Assume that the signal components of the mixture source y are all quantized using the same uniform quantizer. Recalling Equation 4.20, the output noise per sample will then be equal to

$$\sigma_q^2 = \frac{\Delta^2}{12} \tag{4.46}$$

when the quantization interval is Δ, provided the number of quantization levels is high and that no overflows occur [10].

Now define the *rate gain* as the difference between the rate using one optimized entropy coder for the mixture source and the rate when individually optimized entropy coders are applied to each component y_n of the signal.

According to Equation 4.19, the entropy in component y_n is given by

$$H^{(n)}(Y_n) = \frac{1}{2} \log_2[\eta_{y_n} (\frac{\sigma_{y_n}}{\Delta})^2]. \tag{4.47}$$

The average first order entropy per symbol of N samples of the decomposed source using the same Δ for all components is thus

$$
\begin{aligned}
H_N(Y) &= \frac{1}{N} \sum_{n=0}^{N-1} H^{(n)}(Y_n) \\
&= \frac{1}{N} \sum_{n=0}^{N-1} \frac{1}{2} \log_2 \left[\eta_{y_n} (\frac{\sigma_{y_n}}{\Delta})^2 \right] \\
&= \frac{1}{2} \log_2 \eta_{y_g} + \frac{1}{2} \log_2 \left[\prod_{n=0}^{N-1} (\frac{\sigma_{y_n}}{\Delta})^2 \right]^{\frac{1}{N}},
\end{aligned} \tag{4.48}
$$

where η_{y_g} is the geometric mean of the η-factors of the component pdfs.

The rate when quantizing *all* components using only one entropy coder is

$$H(Y) = \frac{1}{2} \log_2 [\eta_y (\frac{\sigma_y}{\Delta})^2]. \tag{4.49}$$

The rate difference between direct coding of the mixture source and the coding of the components individually is thus

$$\Delta H = H(Y) - H_N(Y) = \frac{1}{2} \log_2 \frac{\eta_y}{\eta_{y_g}} + \frac{1}{2} \log_2 \frac{\frac{1}{N} \sum_{n=0}^{N-1} \sigma_{y_n}^2}{[\prod_{n=0}^{N-1} \sigma_{y_n}^2]^{\frac{1}{N}}}. \tag{4.50}$$

Rather than evaluating this result directly, we use the optimal bit allocation as a reference and proceed to compare the above result with that.

Comparison of Bit-Allocation and Entropy Coder Allocation

A comparison between the bit allocation method and the method based on optimal entropy coding of each component can be done in a similar way as in Section 4.1.2. That is, we compute the rate difference between the two methods for equal distortions.

Combining Equations 4.43 and 4.46, and inserting the expression for Δ^2 into Equation 4.48, we get after some manipulations

$$H_N(Y) = b + \frac{1}{2} \log_2 \left(\frac{\eta_{y_g}}{12\epsilon_{y_g}} \right), \tag{4.51}$$

where $b = \frac{B}{N}$. Compared to Equation 4.23, the only difference is that the geometric means of the factors ϵ and η replace the one-component factors. If all the components are governed by the same probability law, which is often a reasonable assumption, the new expression is equal to the one in Equation 4.23. Even in the case of sources with mixture statistics, Equation 4.51 tells us that the arithmetic over geometric means of the variances play the same role for both quantization strategies, as they do not enter the equation.

We refer to Table 4.1 for relevant numerical values for the rate differences for the two systems.

4.3.3 Dynamic Entropy Coder Allocation

In the previous section we have assumed that the signal statistics for each component is known. Assume now that we have so many sources that it would be unrealistic to use one entropy coder per source. This is mostly because it would be necessary to transmit as side information the entropy coder index. The more entropy coders, the longer would the codewords representing the addresses become.

The many components may occur because we decompose the signal in a large number of channels, or because the signal is nonstationary, but may be split in small quasistationary parts with different statistics. The question is now, how do we best select K entropy coders to represent all the different sources, when K is only a fraction of the total number of sources?

This situation is related to dynamic bit allocation, but instead of assigning bits to each component, we instead assign the best available entropy coder. Like in the bit allocation case, the average number of bits generated by the different entropy coders will vary.

To simplify the statistical model we assume fixed pdfs for all the components except for allowing the signal variances to vary. In fact we model the variances by introducing a continuous *variance probability density function* (vpdf). Ideally, we should then use an infinite number of entropy coders to reach optimality for such a case. Experience has shown that a good approximation to the vpdf of subband image signals is the exponential distribution [66].

Using a finite number of entropy coders, the objective will now be to minimize the *coding loss* due to *entropy coder mismatch*. This is done by first selecting the number K of entropy coders to be used. Then the task is to find the K best entropy coders for covering the dynamic range of the vpdf, and to derive the ranges of variances for which each entropy coder should be applied. It turns out that this problem is related to the theory of the pdf-optimization of scalar quantizers as discussed in Section 4.1.1 [66]. The complete theory will not be included here, but the loss curve for a special illustrative case is given in Figure 4.11. In the example the sources are quasistationary with Gaussian pdfs. The variance is distributed as an exponential pdf. In these models it is assumed that uniform quantization is optimal in combination with entropy coding. The same quantizer, i.e. the same value for Δ, is applied everywhere resulting in a fixed noise level.

From Figure 4.11 it is interesting to note that there is a significant loss

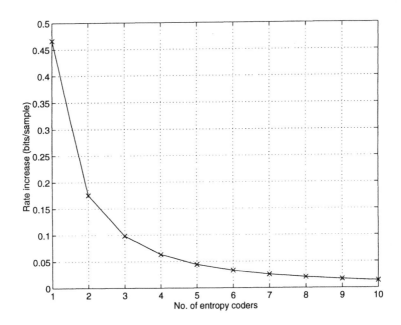

Figure 4.11: *Entropy-increase as a function of the number of entropy coders for a nonstationary Gaussian source with exponential variance distribution.*

if using only one entropy coder, but in a practical case it is not necessary to go beyond 6–8 entropy coders to get close to the limit of zero rate increase as for an infinite number of entropy coders.

The given rate loss depends both on the pdf of the signal and the vpdf. Different results will be obtained if changing any one of the two distributions. But the necessary number of quantizers will be in the same range. A coding example using dynamic entropy coding is included in Chapter 8.

4.4   Summary

This chapter has introduced quantization methods and how to efficiently exploit bit resources to optimize coding of vectors of data containing components with different statistical properties. In the theory of this chapter we have not directly related the components of the mixture source to sub-band coder decomposition. In fact the gains we have defined by using bit allocation or entropy coder allocation have compared direct quantization of the components as a mixture source to the individual quantization of each component. As we shall see in the Chapter 5, the gain formulas change for subband coding because the reference model used when defining gains is the digital representation of the input signal.

One conclusion that can be drawn from this chapter is that entropy coding methods are better than fixed rate quantization based on pdf-optimized quantizers in all cases considered. A second conclusion is that it is of utmost importance to adapt our coding methods to the local statistics of the components. Finally, vector quantization is optimal for stationary sources. But the complexity problem can be prohibitive for many practical systems. The handling of nonstationary signals in VQ systems has not been addressed here, but similar to bit allocation and entropy allocation, it becomes necessary to include some form of codebook allocation [67, 68, 69].

Chapter 5

Gain-Optimized Subband Coders

In this chapter we describe how to optimize subband coders based on mathematical criteria. Bit allocation and entropy coder allocation strategies for mixture sources have been derived in Chapter 4. These results are directly applicable here, but the defined gains are now different because we now want to use coding results for the input signal as a reference, rather than the coding of the mixture source in one quantizer/coder. The second important issue we consider is the performance optimization with respect to the filter bank parameters. Although subband coding of images has been an active research field for the last ten years, there is still no consensus about the filter bank optimization criteria. Experiments indicate that the perfect reconstruction (PR) condition is important, but it does not have to be exactly satisfied. The PR condition can be included in the optimization criteria, or alternatively, we may apply filter banks which structurally guarantee PR, as discussed in Chapter 3. The free parameters could then be optimized for other criteria, such as coding gain, localization in time and frequency for the basis functions, and so on. The ultimate goal is to obtain a visually good coding result for as low a bit rate as possible.

In this chapter we discuss filter banks that maximize the *coding gain*. Two basic theories will be treated, one for systems employing scalar PCM quantization, the second for systems where DPCM quantization is used in each band. In both systems the quantization incorporates adaptivity of bit/entropy coder distributions. We will present optimal coding gains for

155

the systems under discussion, and compare these to traditional transform coder gains of optimal square transforms.

5.1   Gain in Subband Coders

Maximizing the coding gain in a subband coder is equivalent to minimization of its output coding noise for a given bit rate. This requires optimal use of the bit resources either through bit assignment or entropy allocation, as described in Chapter 4. The ability of the filter bank to decorrelate the signal is the second important factor.

More accurately, for a system employing bit allocation, the gain is defined as the ratio between the noise generated when quantizing the input signal in a pdf-optimized scalar quantizer, and the subband coder noise when the two systems operate at the same rate.

To make the necessary statistical calculations, we introduce a mathematical model for one of the channels in the coding system consisting of an analysis bandpass filter, a model of the quantizer noise, and the synthesis bandpass filter of the same channel. The model is shown in Figure 5.1.

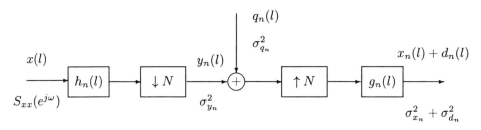

Figure 5.1: *A single analysis-synthesis channel incorporating the additive noise quantizer model.*

In greater detail, the two first blocks represent the decimating analysis filter in channel n, where $\{h_n(l),\ l = 0, 1, \ldots, L_a - 1\}$ is the unit sample response of the bandpass filter, and $\downarrow N$ is the N-times downsampler. We model the quantizer by an additive noise term with sample values $q_n(l)$ and variance $\sigma_{q_n}^2$. The additive noise model is fairly accurate for scalar quantization under the assumption of high bit rate and "sufficient" signal bandwidth. A high number of bits provides a small quantization interval. The necessary bandwidth should provide a signal variation from sample to sample that, on the average, guarantees that different quantization intervals are employed for neighboring samples. In this way there will be little correlation between the noise samples, and it is therefore safe to assume that the noise is white [10]. At low rates a somewhat improved quantizer model includes a signal scaling factor [4].

The next two blocks in the signal chain of Figure 5.1 represent an interpolating synthesis bandpass filter with unit sample response $\{g_n(l), l = 0, 1, \ldots, L_s - 1\}$, and the symbol $\uparrow N$ indicates the introduction of $N - 1$ zero values between the original samples.

The input signal is denoted $x(l)$, and the subband signal in channel n $y_n(l)$. The output signal consists of two components, $x_n(l)$, which is the exact output without quantization of the subband signal (the result of using $y_n(l)$ as the input to the synthesis filter), and $d_n(l)$, which is the output noise component (the result of using $q_n(l)$ as the input to the synthesis filter). The variances of the two output components are $\sigma_{x_n}^2$ and $\sigma_{d_n}^2$, respectively.

In Equation 4.35 we introduced a quantizer noise model valid under the above conditions (repeated here for convenience),

$$\sigma_{q_n}^2 = \epsilon_{y_n} \frac{\sigma_{y_n}^2}{L_n^2}, \tag{5.1}$$

where ϵ_{y_n} is the quantizer performance factor given in Equation 4.12.

In Chapter 4 we compared different ways of quantizing a mixture source, representing e.g. the collection of subband samples. In the bit allocation method, the gain was defined as a relation between the noise from optimal quantization of all the components in one scalar quantizer and the average noise when individually quantizing the components in optimal quantizers applying the "correct" number of bits. As we now are interested in the advantage of using a *subband coder* with inherent bit allocation over a pure scalar quantizer, we have to use as a reference the noise in an optimal scalar quantization applied to the *input* signal. It is then clear that the filtering operations both in the analysis and synthesis filter banks will influence the resulting gain.

The reference noise can be calculated from

$$\sigma_{PCM}^2 = \epsilon_x \frac{\sigma_x^2}{L_{y_g}^2}, \tag{5.2}$$

where the number of levels is chosen to L_{y_g} to make the comparison fair[1].

The subband coder output noise variance will be a function of the input signal statistics, the quantizer parameters (L_n and ϵ_{y_n}), and the filter parameters that we want to optimize.

There are two main differences from the theory in Chapter 4:

1. The variance of the input signal will in general be different from the variance of the components: $\sigma_x^2 \neq \sigma_y^2 = \frac{1}{N} \sum_{n=0}^{N-1} \sigma_{y_n}^2$. Equality occurs if unitary systems are applied.

2. As the reference is now the input signal, the mixture distribution must be replaced by the input pdf. This means that $\epsilon_x \neq \epsilon_y$ and $\eta_x \neq \eta_y$.

Review of Filtering Operations on Stochastic Signals

Figure 5.2 depicts the relevant entities of this section.

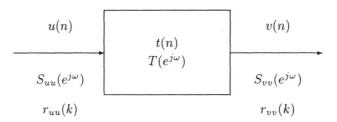

Figure 5.2: *Filtering operations on stochastic signal.*

With a zero-mean signal $u(n)$ characterized by its *power spectral density* (psd) $S_{uu}(e^{j\omega})$ as input signal and a filter with frequency response $T(e^{j\omega})$, the power spectral density of the output signal is given by

$$S_{vv}(e^{j\omega}) = |T(e^{j\omega})|^2 S_{uu}(e^{j\omega}). \qquad (5.3)$$

Because the variance is the integral of the power spectral density, the output variance becomes

$$\sigma_v^2 = \int_{-\pi}^{\pi} S_{uu}(e^{j\omega})|T(e^{j\omega})|^2 \frac{d\omega}{2\pi}. \qquad (5.4)$$

---

[1] L_{y_g} is the geometric mean of the levels in the different bands

For optimization and interpretation of various results, it is convenient to express the above relation in several different time domain forms, as well. The corresponding time domain equation is

$$\sigma_v^2 = \sum_{l=-\infty}^{\infty} r_{uu}(l) \sum_{i=-\infty}^{\infty} t(i)t(l+i), \qquad (5.5)$$

where $r_{uu}(l)$ is the *autocorrelation function* (acf) of the input signal, which is related to the power spectral density through the inverse Fourier transform, and $t(l)$ is the unit sample response of the filter.

A matrix formalism results from interchanging the order of summations in Equation 5.5, followed by introduction of a new summation variable $k = l + i$,

$$\sigma_v^2 = \sum_{i=-\infty}^{\infty} t(i) \sum_{k=-\infty}^{\infty} r_{uu}(k-i)t(k) = \mathbf{t}^T \mathbf{R}_{uu} \mathbf{t}, \qquad (5.6)$$

where $\mathbf{t} = [\dots, t(-1), t(0), t(1), \dots]^T$ is a column vector consisting of the samples of the unit sample response of the filter, and $\mathbf{R}_{uu}$ is the infinite correlation matrix of the input signal with matrix elements $R_{uu}(k, l) = r_{uu}(k-l)$.

Variance Calculations

In the following derivation we express gain and bit allocation formulas using the formalism in Equation 5.6, but the final results will be put in different forms depending on the application at hand.

The subband signal variance in channel no. n is

$$\sigma_{y_n}^2 = \mathbf{h}_n^T \mathbf{R}_{xx} \mathbf{h}_n. \qquad (5.7)$$

The relation between the output noise and the quantization noise is computed similarly, but because the quantization noise is white, the auto-correlation matrix is diagonal with equal diagonal elements given by the quantization noise variance. Then,

$$\sigma_{d_n}^2 = \frac{\sigma_{q_n}^2}{N} \mathbf{g}_n^T \mathbf{g}_n, \qquad (5.8)$$

where the factor $1/N$ comes from the upsampling operation in which the power is reduced by a factor of N due to the insertion of zero-valued samples.

Combining Equations 5.1, 5.7, and 5.8 we obtain the sought output noise variance in channel n as

$$\sigma_{d_n}^2 = \frac{\epsilon_{y_n} \sigma_{y_n}^2}{L_n^2 N} (\mathbf{g}_n^T \mathbf{g}_n). \tag{5.9}$$

A simplification is obtained by requiring that the synthesis filters be *normalized*:

$$\mathbf{g}_n^T \mathbf{g}_n = \sum_{i=-\infty}^{\infty} g_n^2(i) = \int_{-\pi}^{\pi} |G_n(e^{j\omega})|^2 \frac{d\omega}{2\pi} = 1. \tag{5.10}$$

This assumption does not sacrifice generality, but the analysis filters have to be rescaled accordingly such that perfect reconstruction is maintained in the back-to-back configuration of the analysis and synthesis filter banks.

Using the above results, the total output noise, assuming independent noise contributions, is given by

$$\sigma_d^2 = \sum_{n=0}^{N-1} \sigma_{d_n}^2 = \sum_{n=0}^{N-1} \frac{\epsilon_{y_n} \sigma_{y_n}^2}{L_n^2 N}. \tag{5.11}$$

The obtained output noise equation shows that the subband coder output noise is, in fact, equal to the noise when quantizing the subband components. The reason is the constraint of unit norm synthesis filters and white quantization noise. The gain formulas here will therefore differ from the gain formulas in Chapter 4 only because the reference signal is different.

Resulting Formulas

For general subband decomposition, provided the synthesis filter bank is normalized according to Equation 5.10, we compute the coding gain using bit allocation following the same procedure as in Section 4.3.1. Assuming white, independent noise in all channels, we can still define the average distortion like in Equation 4.31, but now with the understanding that this is the noise variance *in the reconstructed signal*:

$$\sigma_d^2 = \frac{1}{N} \sum_{n=0}^{N-1} \sigma_{q_n}^2. \tag{5.12}$$

The optimization follows Equations 4.32 to 4.43, the last equation repeated here for convenience:

$$\sigma_d^2 = \frac{1}{N} \sum_{n=0}^{N-1} \sigma_{q_n}^2 = \epsilon_{y_g} \frac{\sigma_{y_g}^2}{L_g^2}, \tag{5.13}$$

where the subindex g denotes the geometric mean of the respective entities. This is an expression for the reconstruction noise variance for a subband coding system when using optimal bit allocation. The corresponding noise variance when using scalar quantization directly on the input signal x is

$$\sigma_x^2 = \epsilon_x \frac{\sigma_x^2}{L_g^2} \tag{5.14}$$

and the subband coding gain is found as

$$G_{SBC} = \frac{\sigma_x^2}{\sigma_d^2} = \frac{\epsilon_x}{\epsilon_{y_g}} \frac{\sigma_x^2}{\left[\prod_{n=0}^{N-1} \sigma_{y_n}^2\right]^{\frac{1}{N}}}. \tag{5.15}$$

In the formula above, the subband variances can be calculated from any of the forms:

$$\sigma_{y_n}^2 = \sum_{l=-\infty}^{\infty} r_{xx}(l) \sum_{i=-\infty}^{\infty} h_n(i) h_n(l+i) \tag{5.16}$$

$$= \mathbf{h}_n^T \mathbf{R}_{xx} \mathbf{h}_n \tag{5.17}$$

$$= \int_{-\pi}^{\pi} S_{xx}(e^{j\omega}) |H_n(e^{j\omega})|^2 \frac{d\omega}{2\pi}, \tag{5.18}$$

provided the unit sample synthesis filter responses are normalized according to Equation 5.10. We repeat the bit/level allocation formula from Equation 4.42, and stress that the above gain formula is valid only when using the appropriate number of bits from this formula.

$$b_i = \log_2(L_i) = \frac{B}{N} + \frac{1}{2} \log_2 \left(\frac{\epsilon_{y_i} \sigma_{y_i}^2}{\epsilon_{y_g} \sigma_{y_g}^2} \right), \qquad i = 0, 1, \ldots, N-1. \tag{5.19}$$

The entropy gain in a subband coder when using scalar quantization and correct entropy coders for each component over direct entropy coding of the input signal, can be obtained by comparing the above derivation and

Equation 4.50. Basically we replace $\sigma_y^2 = \frac{1}{N}\sum_{n=0}^{N-1}\sigma_{y_n}^2$ by σ_x^2, and ϵ_y by ϵ_x in Equation 4.50. The resulting rate gain is

$$\Delta H = H(X) - H_N(Y) = \frac{1}{2}\log_2\frac{\eta_x}{\eta_{y_g}} + \frac{1}{2}\log_2\frac{\sigma_x^2}{[\prod_{n=0}^{N-1}\sigma_{y_n}^2]^{\frac{1}{N}}}. \qquad (5.20)$$

Comments

In these derivations we have deliberately avoided to constrain the filter banks, except for normalization on the synthesis side, and a corresponding inverse scaling on the analysis side. In a practical system we need some further interrelations between the filters to guarantee perfect or near perfect reconstruction in the absence of quantization noise, and it is also obvious that the different unit sample responses must represent some frequency partitioning of the signal bandwidth.

In practice, for nonstationary signals with unknown statistics, we need estimates of the local subband variances. In coder implementations we use dynamic bit allocation to adapt to the local statistics of the images. Likewise, the entropy coder allocation must also be based on local statistics. The estimates are therefore based on the subband signals in a small local area around the samples to be quantized. As we shall see later in this chapter, the filter bank should also rely on the local statistics. Adaptivity in filter banks is a more involved subject which will not be pursued in the present chapter, although the presented theory can allow for adaptivity.

It is obvious that the ratios $\epsilon_x/\epsilon_{y_g}$ and η_x/η_{y_g} in the two gain formulas of this chapter are different from the corresponding values in the formulas in Chapter 4, Equations 4.45 and 4.50. If the input signal is Gaussian, so will the subband signals be, irrespective of the filter coefficient choice. Then the above ratios are equal to unity. In all other cases the pdf will change during filtering, and the ratios become different from unity.

An example might clarify the role of the constants ϵ and η. Assume that the input signal is uniformly distributed [10]. Depending on the decomposition process, the signal will tend towards a Gaussian distribution. In Table 5.1, we give values when the components are either Laplacian or Gaussian distributed. When all the components have the same distribution, so will the geometrical mean. It is interesting to observe that the ratios $\epsilon_x/\epsilon_{y_g}$ for the Laplacian and Gaussian cases are different.

Relation	Laplacian	Gaussian
$\epsilon_x/\epsilon_{y_g}$	0.22	0.37
η_x/η_{y_g}	0.81	0.70

Table 5.1: *Gain factors for bit allocation and entropy coding of a uniform pdf input. The subband signals are supposed to be either Laplacian or Gaussian.*

We stress that the comparison made in Equation 4.51 between the necessary bit rates using bit allocation and entropy coder allocation for systems producing equal distortion is still valid. This implies that SBC systems employing entropy coder allocation always outperform similar systems using bit allocation. However, have in mind the unavoidable bit variations associated with entropy coding.

The gain limit under perfect reconstruction conditions and stationary signals will be the topic of the next subsections.

5.1.1 Optimal Square Transforms

Both gain formulas from the previous section depend on the input signal statistics and the analysis filter bank. For a given signal statistic it should therefore be possible to maximize the gain by selecting appropriate filters.

Before going into the optimization, we briefly review the theory for optimal transforms in transform coding, which is a special filter bank case. It is well known that the optimal transform is the so-called *Karhunen-Loève transform* (KLT) [70, 71]. The KLT is composed of the *eigenvectors* of the autocorrelation matrix found from the equation

$$\mathbf{R}_{xx}\mathbf{h}_n = \lambda_n \mathbf{h}_n. \tag{5.21}$$

If the eigenvectors are column vectors, the KLT matrix is composed of the eigenvectors as its rows:

$$\mathbf{K} = [\mathbf{h}_0 \mathbf{h}_1 \dots \mathbf{h}_{N-1}]^T. \tag{5.22}$$

Because the rows of this matrix are the eigenvectors of an autocorrelation matrix, the Karhunen-Loève transform can always be made unitary. The rows are automatically mutually orthogonal when $\mathbf{R}_{xx}$ is *Toeplitz*, as in this case [23]. Furthermore, normalization follows from the fact that the synthesis responses are normalized. By performing the transform on the input vector $\mathbf{x}$, we get the *transform coefficients*

$$\mathbf{y} = \mathbf{K}\mathbf{x}. \tag{5.23}$$

The inverse in the unitary case is given by the Hermitian of the transform matrix and can therefore be found from

$$\mathbf{x} = \mathbf{K}^H \mathbf{y}. \tag{5.24}$$

Let us consider the optimality of the Karhunen-Loève transform from our gain formula in Equation 5.15, which is general, and thus should also be applicable for the square transform case.

To find $\sigma_{y_n}^2$ we use the expression in Equation 5.17 and the eigenvalue equation, Equation 5.21:

$$\sigma_{y_n}^2 = \mathbf{h}_n^T \mathbf{R}_{xx} \mathbf{h}_n = \mathbf{h}_n^T \lambda_n \mathbf{h}_n = \lambda_n. \tag{5.25}$$

The relation $\mathbf{h}_n^T \mathbf{h}_n = 1$ is obtained because the system is unitary and therefore both the analysis and the synthesis transforms must have unit norm. The optimal transform coder gain is thus

$$G_{TC} = \frac{\epsilon_x}{\epsilon_{y_g}} \frac{\sigma_x^2}{\left[\prod_{n=0}^{N-1} \lambda_n\right]^{\frac{1}{N}}}. \tag{5.26}$$

which is in accordance with the standard formula.

5.1.2　Optimal Nonsquare Transforms (Filter Banks)

Obviously, there are more degrees of freedom in a nonsquare transform as compared to a square transform with the same number of channels, N. We therefore approach the challenge on how to exploit these extra degrees of freedom. If a filter length larger than N can give improved performance, we can expect the performance of the subband coder to lie in between a transform coder and an *ideal predictive coder*, because the latter can remove all redundancy due to correlation.

In the following we will derive the optimal filter banks for the *unitary* and the general *nonunitary* cases and compare these to the Karhunen-Loève square transform.

A necessary condition for optimal gain performance is complete *decorrelation* between the subbands. This is reasonable from the point of view that the same information should not be represented more than once. This would not be the case when two or more channels are correlated. From a mathematical point of view, the gain in Equation 5.15 is maximized by making the subband variances small. The variance expression in Equation 5.18 indicates that the variance is minimized when the bandwidth of $|H_n(e^{j\omega})|^2$ is minimized. However, the filters have to satisfy the PR condition, which limits the minimum bandwidth. It is plausible that the minimum is reached for ideal filters with perfect frequency separation, that is, contiguous and nonoverlapping filters that cover the complete frequency range $[-\pi, \pi]$.

Decorrelation in Perfect Filter Banks

The perfect decorrelation property of nonoverlapping filters can be found from the crosscorrelation between samples at different positions in two channels:

$$
\begin{aligned}
r_{y_i,y_j}(n, m) &= E[y_i(n)y_j(m)] \\
&= E\left[\sum_k h_i(k)x(n-k)\sum_l h_j(l)x(m-l)\right] \\
&= \sum_k\sum_l h_i(k)h_j(l)E[x(n-k)x(m-l)] \\
&= \sum_r r_{xx}(n-m+r)\sum_l h_i(l-r)h_j(l),
\end{aligned}
\tag{5.27}
$$

where the last equality has been obtained by introducing the substitution $k = l - r$ and summing over r instead of k. Parseval's relation [72] allows the crosscorrelation to be expressed in terms of frequency domain filter characterization:

$$
r_{y_i,y_j}(n, m) = \sum_r r_{xx}(n-m+r)\int_{-\pi}^{\pi} H_i(e^{-j\omega})H_j(e^{j\omega})e^{j\omega r}\frac{d\omega}{2\pi}.
\tag{5.28}
$$

From this expression it is obvious that the integral is equal to zero if the two filter responses do not overlap. This proves that the outputs from nonoverlapping filters are completely decorrelated irrespective of the correlation properties of the input signal.

Using the idealized filters above, perfect reconstruction is obtained by imposing the following relation between the analysis filter $H_n(e^{j\omega})$ and the corresponding synthesis filter $G_n(e^{j\omega})$:

$$|H_n(e^{j\omega})| = \begin{cases} \frac{N}{|G_n(e^{j\omega})|} & \text{for } \omega \in \pm[\frac{\pi n}{N}, \frac{\pi(n+1)}{N}] \\ 0 & \text{otherwise,} \end{cases}, \quad n = 0, 1, \ldots, N-1, \quad (5.29)$$

where the factor N must be included to compensate for the power loss during interpolation.

Optimal Unitary Filter Banks

In traditional filter bank theory of subband coders, it is implicitly assumed that optimality is reached by employing ideal brick-wall filters (see page ix) [10]. For this filter type the analysis and synthesis filter banks must have the same frequency response magnitudes, thus in channel n we have according to Equation 5.29

$$|H_n(e^{j\omega})|^2 = \begin{cases} N & \text{for } \omega \in \pm[\frac{\pi n}{N}, \frac{\pi(n+1)}{N}] \\ 0 & \text{otherwise.} \end{cases}, \quad n = 0, 1, \ldots, N-1. \quad (5.30)$$

Furthermore, if the filters have nonlinear phase, the two filters must have opposite phase responses to obtain perfect reconstruction. The phases do not enter the gain formulas, so we do not worry about them at the moment. Thus, in a unitary filter bank the unit sample responses in the analysis and synthesis filter banks are reversed versions of each other. With the above analysis filters the gain in the "ideal" subband coder is given by

$$\begin{aligned} G_{SBC,ideal} &= \frac{\epsilon_x}{\epsilon_{y_g}} \frac{\sigma_x^2}{[\prod_{n=0}^{N-1} \sigma_{y_n}^2]^{\frac{1}{N}}} \\ &= \frac{1}{N} \frac{\epsilon_x}{\epsilon_{y_g}} \left[\prod_{n=0}^{N-1} 2 \int_{\frac{\pi n}{N}}^{\frac{\pi(n+1)}{N}} \frac{S_{xx}(e^{j\omega})}{\sigma_x^2} \frac{d\omega}{2\pi} \right]^{-\frac{1}{N}}, \quad (5.31) \end{aligned}$$

where the factor 2 is due to the integration interval which is here only taken over positive frequencies.

The prerequisite for the gain of the synthesis filters – i.e. the synthesis filter normalization – can be checked as follows:

$$\int_{-\pi}^{\pi} |G_n(e^{j\omega})|^2 \frac{d\omega}{2\pi} = 2 \int_{\frac{\pi n}{N}}^{\frac{\pi(n+1)}{N}} N \frac{d\omega}{2\pi} = 1. \quad (5.32)$$

The obtained gain formula depends only on the input spectrum relative to the input power, the quantizer performance factors, and the frequency partitioning.

The coding gain limit, as the number of channels increase to infinity, can be calculated as follows: The gain in Equation 5.31 is rewritten as

$$G_{SBC,ideal} = \frac{1}{N} \frac{\epsilon_x}{\epsilon_{y_g}} \exp\left\{-\frac{1}{N} \sum_{n=0}^{N-1} \ln\left(2 \int_{\frac{\pi n}{N}}^{\frac{\pi(n+1)}{N}} \frac{S_{xx}(e^{j\omega})}{\sigma_x^2} \frac{d\omega}{2\pi}\right)\right\}. \quad (5.33)$$

The bandwidth is $\frac{\pi}{N}$ in all integrals. When $N \to \infty$ we set $d\omega \approx \Delta\omega = \frac{\pi}{N}$. Furthermore, within each band we replace the continuous frequency by the center frequency, here denoted by ω_n.

Inserting this result in Equation 5.33, we obtain

$$G_{SBC,ideal} \approx \frac{1}{N} \frac{\epsilon_x}{\epsilon_{y_g}} \exp\left\{-\frac{1}{N} \sum_{n=0}^{N-1} \ln\left(\frac{S_{xx}(e^{j\omega_n})}{\sigma_x^2} \frac{1}{N}\right)\right\}$$

$$= \frac{\epsilon_x}{\epsilon_{y_g}} \sigma_x^2 \exp\left\{-\sum_{n=0}^{N-1} \ln(S_{xx}(e^{j\omega_n})) \frac{1}{N}\right\}$$

$$= \frac{\epsilon_x}{\epsilon_{y_g}} \sigma_x^2 \exp\left\{-2 \sum_{n=0}^{N-1} \ln(S_{xx}(e^{j\omega_n})) \frac{\Delta\omega}{N} \frac{N}{2\pi}\right\}$$

$$\approx \frac{\epsilon_x}{\epsilon_{y_g}} \sigma_x^2 \exp\left\{-\int_{-\pi}^{\pi} \ln S_{xx}(e^{j\omega}) \frac{d\omega}{2\pi}\right\} = \frac{\epsilon_x}{\epsilon_{y_g}} \gamma_x^{-2}, \quad (5.34)$$

which is the inverse of the *spectral flatness measure* times the ratio between the quantizer performance factors. For Gaussian signals, where the quantizer performance factor ratio is unity, maximum gain, attainable by utilizing second order statistics of a process, is obtained [10].

Optimal Nonunitary Case

To improve the performance of a subband coder for a finite number of channels, we have to allow for nonunitary filter banks. That is, we can use different filter magnitudes for the analysis and the corresponding synthesis filters. This can be motivated in several ways. As nonunitary filter banks have been recognized as a viable alternative only by a few researchers up to now, [73, 36, 74] we will use two different motivating arguments for this before we present a more rigorous mathematical treatment.

The first argument is the following: In subband coders with a finite number of brick-wall filters, there is no mechanism for removing the *intra-band* correlation. Intra-band decorrelation can be obtained by employing some sort of whitening in each subband before quantization. The whitening process would have to be counteracted in the synthesis filter bank to obtain perfect reconstruction, thus requiring different filters in the analysis and synthesis filter banks.

An alternative argument is based on an observation of the gain formula in Equation 5.15. If employing unitary filter banks, we have learned in Chapter 2 that the unit sample responses in an analysis filters, as well as in a synthesis filters, have to be orthogonal to shifted versions of themselves by lN samples, where l is an integer. This means that certain lags in the autocorrelation function do not play any role in the optimization because the corresponding *deterministic filter correlation function* of the filter

$$\rho_{h_n}(lN) = \sum_{i=-\infty}^{\infty} h_n(i)h_n(lN+i) = \delta(l), \qquad (5.35)$$

is zero. This may at first glance seem to be perfectly OK when minimizing the denominator of Equation 5.15. A closer analysis reveals that this does not necessarily make each factor in the product smaller, but reduces the possibility of adapting the filter coefficients to the complete autocorrelation function.

The zero terms appear for every N lags of the autocorrelation function. We would therefore expect the suboptimality to be more pronounced for a unitary filter bank with few channels, and to decay as the number of channels increases.

It is worth noticing that this mechanism does not influence the optimality of a unitary square transform because the filter lengths are all equal to N, in which case there will be no overlap between shifted-by-N versions of the filters and certainly not for delays of lN in Equation 5.35.

Now, let us consider the theoretically optimal N-channel filter bank. The frequency domain version of the subband variances, Equation 5.18, is for this purpose inserted in Equation 5.15, and the relation between analysis and synthesis filters in Equation 5.29 is used to replace $|H_n(e^{j\omega})|$ by $N/|G_n(e^{j\omega})|$. With the above requirement, the gain formula in Equation 5.31

can be reformulated as

$$G_{SBC} = \frac{1}{N^2} \frac{\epsilon_x}{\epsilon_{y_g}} \left[\prod_{n=0}^{N-1} 2 \int_{\frac{\pi n}{N}}^{\frac{\pi(n+1)}{N}} \frac{S_{xx}(e^{j\omega})}{\sigma_x^2} |G_n(e^{j\omega})|^{-2} \frac{d\omega}{2\pi} \right]^{-\frac{1}{N}}. \qquad (5.36)$$

Because all channels are uncorrelated, we can minimize each term in the product by the use of Schwartz' inequality, usually expressed as

$$|\langle \alpha_n, \beta_n \rangle|^2 \leq \| \alpha_n \|^2 \| \beta_n \|^2, \qquad (5.37)$$

where $\langle \cdot, \cdot \rangle$ signifies the inner product defined by

$$\langle \alpha_n, \beta_n \rangle = 2 \int_{\frac{\pi n}{N}}^{\frac{\pi(n+1)}{N}} \alpha_n(\omega) \beta_n(\omega) \frac{d\omega}{2\pi}, \qquad (5.38)$$

and $\| \cdot \|$ is the norm operator.

Setting

$$\alpha_n = \sqrt{\frac{S_{xx}(e^{j\omega})}{\sigma_x^2}} |G_n(e^{j\omega})|^{-1} \qquad (5.39)$$

$$\beta_n = |G_n(e^{j\omega})|, \qquad (5.40)$$

so that $\| \beta_n \|^2 = 1$, direct application of the Schwartz inequality, where the right-hand side of Equation 5.37 is used for the integrand, gives the upper limit for the coding gain:

$$G_{SBC} \leq \frac{1}{N^2} \frac{\epsilon_x}{\epsilon_{y_g}} \left[\prod_{n=0}^{N-1} 2 \int_{\frac{\pi n}{N}}^{\frac{\pi(n+1)}{N}} \sqrt{\frac{S_{xx}(e^{j\omega})}{\sigma_x^2}} \frac{d\omega}{2\pi} \right]^{-\frac{2}{N}}. \qquad (5.41)$$

The maximum is attained for equality, which is obtained for

$$\sqrt{\frac{S_{xx}(e^{j\omega})}{\sigma_x^2}} |G_n(e^{j\omega})|^{-1} = c_1 |G_n(e^{j\omega})|, \text{ for } \omega \in \pm[\tfrac{\pi n}{N}, \tfrac{\pi(n+1)}{N}], \qquad (5.42)$$

where c_1 is an arbitrary constant. Again, using Equation 5.29, we can solve for the frequency response of analysis filter no. n:

$$|H_n(e^{j\omega})| = \begin{cases} c_2 \left[\frac{S_{xx}(e^{j\omega})}{\sigma_x^2} \right]^{-1/4} & \text{for } \omega \in \pm[\tfrac{\pi n}{N}, \tfrac{\pi(n+1)}{N}] \\ 0 & \text{otherwise,} \end{cases} \qquad (5.43)$$

where c_2 is another arbitrary constant. $H_n(z)$ is a so-called *half-whitening* filter.

With the above selection of filters the maximum gain in subband coders with N channels is given by

$$G_{SBC,opt} = \frac{1}{N^2} \frac{\epsilon_x}{\epsilon_{y_g}} \left[\prod_{n=0}^{N-1} 2 \int_{\frac{\pi n}{N}}^{\frac{\pi(n+1)}{N}} \sqrt{\frac{S_{xx}(e^{j\omega})}{\sigma_x^2}} \frac{d\omega}{2\pi} \right]^{-\frac{2}{N}}. \tag{5.44}$$

The theoretical gain limit as $N \to \infty$ is also in this case given by the inverse of the spectral flatness measure times the ratio between the quantizer performance factors. The derivation may follow the same lines as for the unitary case.

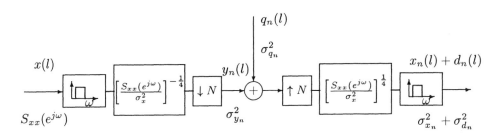

Figure 5.3: *Interpretation of one channel of an optimal subband coder with scalar quantizer.*

To summarize, we stress that the actual filter operations consist of a complete inter-band decorrelation by ideal frequency separation among the bands, then a half-whitening filtering within each band [10]. This can be illustrated as in Figure 5.3.

An interpretation of the structure in Figure 5.3 is the following: Within the frequency region defined by the ideal bandpass filter, the coding is equivalent to open-loop differential coding (see Chapter 1), for which the half-whitening operation is the optimal predictor [10]. We therefore also refer to this filter as an open-loop filter. In the reconstruction part of the system the inverse filter must be used to recreate the spectral shape of the channel.

Another important result is that the phase of the channel filters do not enter in the optimization and is therefore irrelevant for gain optimization.

5.1.3 Theoretical Coding Gains

In the previous sections we have derived the gains for three different "optimal" filter banks for subband coding:

- Optimal square transform = Karhunen-Loève transform (unitary).

- The ideal brick-wall filter bank (unitary).

- Psd-optimized filter bank (nonunitary).

In this section we calculate theoretical gains for a given input power spectral density. The statistical model we apply for the input signal is the much used AR(1)-model with autocorrelation function $r_{xx}(k) = \rho^{|k|}\sigma_x^2$, where $\rho = 0.95$, and also assume that the signal is Gaussian. For Gaussian signals we recall that $\epsilon_x/\epsilon_{y_g} = 1$.

It is possible to find the gain of the Karhunen-Loève transform for the given process explicitly as given in [10]:

$$G_{KLT} = (1 - \rho^2)^{-\frac{N-1}{N}}. \tag{5.45}$$

The gains for the three systems are shown as functions of the number of channels in Figure 5.4.

Note the performance advantage of the optimal filter bank for a small number of subbands. For a large number of subbands the differences in coding gain between the two filter banks become negligible. This is due to the upper limit of gain achievable by exploitation of second order statistics. For the chosen input process, the maximum gain is 10.11 dB [10]. All curves approach this limit as the number of channels is increased, but the Karhunen-Loève transform clearly has inferior performance due to the short unit pulse responses. Remember, both filter banks have infinite length unit sample responses in all channels!

5.1.4 Design Examples

Due to the assumption of ideal frequency separation, the above discussion on optimal nonsquare transforms (filter banks) is valid for infinite length filters only. In practice we need fairly short filters due to the nonstationarity of the signals to be coded. The question is then, how well does an optimized filter bank with *finite* length responses perform?

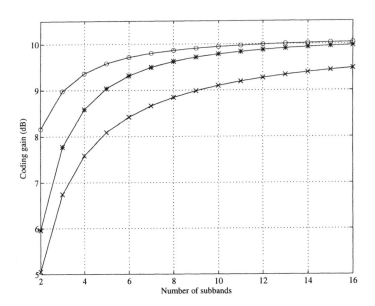

Figure 5.4: *Theoretical coding gains for: Optimal nonunitary filter bank (○),*
brick-wall filter bank (∗), and Karhunen-Loève transform (×).

Two-channel Examples

We now consider a subband coder with a two-channel filter bank consisting of
nonunitary, linear phase filters, and optimize for an AR(1), $\rho = 0.95$ process.
The filter banks considered in [73] were all relatively short (up to 9 taps)
to allow for straightforward optimization. Figure 5.5 shows the frequency
response of the analysis filters of length 5 and 3 unit sample responses for
the lowpass and highpass filters, respectively [73].

Observe the characteristic lift in the lowpass response. The lift results
in a flatter spectrum for the lowpass channel signal, and consequently a less
correlated signal in that band.

Two other two-channel examples were shown in Figures 3.34 and 3.35,
respectively. These will in the following be called *lin6* and *lin8*, respectively.

With the gradient search method proposed by Nayebi et al. [20, 21], and
modified as presented in [36] and in Section 2.5 of this book, we can optimize
for filters with long responses. Coding gains for filter lengths 16, 32, 128,

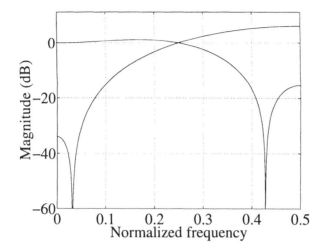

Figure 5.5: *Analysis frequency responses for 2-channel filter bank [73].*

and an infinite no. of taps are shown in Table 5.2.

We observe that the gain increases with the filter length, as expected. The new filter banks have considerably larger gain than the unitary f16B filters by Johnston [17] as well as in the simple filter bank above (Figure 5.5).

The frequency responses of the new filter banks are compared to that of the optimal half-whitening filters for an AR(1), $\rho = 0.95$ process. In Figure 5.6 both analysis and synthesis responses for the three new filter banks are shown. The response of the optimal half-whitening filter is plotted in dotted lines and normalized such that it can be compared to the optimized lowpass responses.

The 32-tap lowpass filters have responses close to the corresponding optimal open-loop DPCM filters for frequencies less than 0.25. This is also the case for the highpass filters in the 0.25 to 0.50 frequency range, but due to the scaling according to Equation 5.10 there is a fixed distance between corresponding curves. For the 128-tap filter bank the filters are practically identical to the optimal open-loop DPCM filters. In addition we observe the improved transition width at the half band frequency, thus aliasing due to downsampling is reduced. These responses clearly confirm that the improved coding gains of nonunitary filter banks are due to the possibility of having optimal open-loop filters in each filter bank channel.

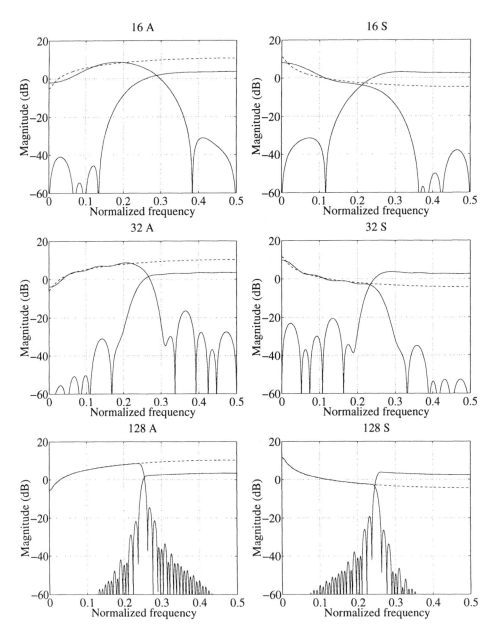

Figure 5.6: *Comparison of obtained filter banks and optimal half-whitening filters: A=analysis and S=synthesis.*

Filter bank	Coding gain (dB)
f16B [17]	5.91
5,3-tap [73]	6.31
lin6	6.21
lin8	6.38
16-tap	7.72
32-tap	8.06
128-tap	8.160
optimal	8.161

Table 5.2: *AR(1), $\rho = 0.95$ coding gain for two-channel filter banks.*

Eight-Channel System

We next proceed to optimize a uniform filter bank with 8 channels and finite length responses, here specified to 16 taps. Figures 5.7 and 5.8 depict the analysis and synthesis frequency responses for this filter bank.

In this example the half-whitening mechanism presented earlier is not totally justified, since the filter bank exhibits severe aliasing. Similar to the SSKF filter bank [73], we can detect a small lift in the analysis lowpass response, compensated for by a decaying lowpass synthesis response.

For the higher bands there is little difference between corresponding analysis/synthesis filters in the passband regions. We attribute this to the relatively flat response of the AR(1), $\rho = 0.95$ power spectrum at high frequencies. Note the differences between the bandpass and highpass analysis and synthesis responses at low frequencies. The analysis bandpass responses have better attenuation at the, presumed high-energy, lowpass spectral area, whereas these characteristics are not as pronounced for the synthesis responses. In spite of overlapping frequency responses, with associated aliasing, the obtained filter bank has a coding gain of 9.63 dB. This is in fact slightly above the gain offered by the corresponding $N = 8$ brick-wall uniform filter bank (9.62 dB) [75].

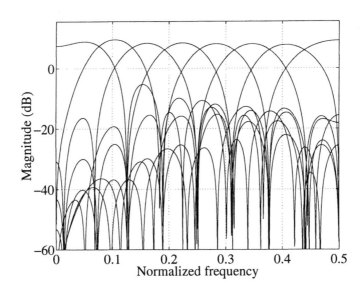

Figure 5.7: *Analysis frequency responses for 8-channel, 16-tap filter bank.*

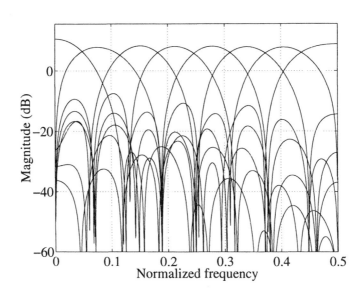

Figure 5.8: *Synthesis frequency responses for eight-channel, 16-tap filter bank.*

5.2 Reformulation of the Optimal Subband System

The optimal subband filters of infinite length provide ideal frequency partitioning and thus inter-band decorrelation. Based on this observation it is easy to change the complete filter bank structure into something simpler. The filtering in one channel can be viewed as a combination of a brick-wall filter and a half-whitening filter as shown in Figure 5.3. As both are linear filters, we can interchange their order. The half-whitening filters have a region of support defined by the subsequent brick-wall bandpass filter. This means that their frequency response outside this area is irrelevant. We therefore extend the region of support of every half-whitening filter to the complete frequency range of the subband coder. Then each branch in the coder contains the same half-whitening filter as shown in Figure 5.9, and each filter has the same input signal.

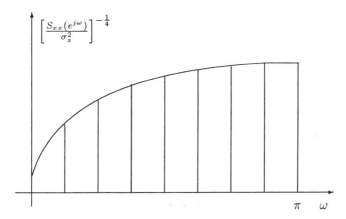

Figure 5.9: *Frequency responses of the collection of bandpass filters in an optimal subband coder after special selection of half-whitening amplification factors.*

Hence, these filters can be pulled out of the branches and combined as shown in Figure 5.10.

We can justify the obtained structure in another way. Consider the optimal open-loop DPCM system in Figure 5.11. Optimality of such a system is obtained if the predictor performs half-whitening [10]. All our theoretical developments are based on the use of a sufficient number of bits that whitens the quantization noise. The optimality of the open-loop DPCM system is

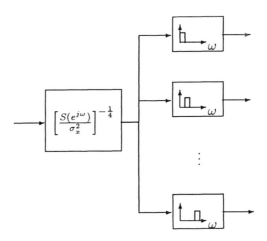

Figure 5.10: *Optimal subband coder system based on fullband half-whitening filter and brick-wall filter bank.*

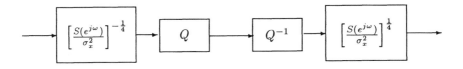

Figure 5.11: *Optimal open-loop DPCM system*

no exception. In a subband coder with high bit rate and *brick-wall* filters, the output coder noise is also going to be white, but lower than in a PCM quantizer. Thus the subband encoder-decoder system with "ideal filters" can be viewed as another scalar quantizer, and we can therefore replace the quantizer in Figure 5.10 with a subband coder with spectral-adaptive use of the bit resources. This is also optimal in terms of coding gain because open-loop DPCM is the best we can do with this kind of structure. However, the output noise in the receiver is nonwhite for this system, shaped by the inverse of the half-whitening filter.

It may be of considerable practical interest to use a structure as illustrated in Figure 5.10. In adaptive systems it may be simpler to do the adaptation in the fullband predictor than in N subband predictors, especially if the input signal can be modeled as a simple AR(1) process.

5.3 System Quantizer replaced by DPCM Coder

In all theory presented so far in this chapter, we have assumed that the quantizer is a stand alone scalar quantizer. To compensate for this we were forced to apply half-whitening in the subbands prior to quantization. This, in turn, implies that the output noise will be colored due to the inverse filtering in the decoder.

However, it is well known that closed-loop DPCM performs better than open-loop DPCM and also produces white output noise [10]. Whether the whiteness property is subjectively advantageous or not, is not entirely clear, but as we are here searching for gain optimality of subband coders, the DPCM option is clearly justified. Assuming that we are able to do optimal bit allocation, and use the optimal predictors in each band, the gain will be equal to the gain in a one-band closed-loop DPCM coder irrespective the number of bands, when disregarding the quantizer performance factors which may influence this statement for non-Gaussian sources.

Even though the performance of such a system seems quite clear, we still need to find the relevant bit allocation formula. A DPCM coder can be viewed as a quantizer with a better performance than a PCM quantizer due to the prediction gain. We can therefore model the new quantizer and replace this model with the quantizer performance in Equation 5.1 for PCM as

$$\sigma_{q_n}^2 = \epsilon_{d_n} \gamma_{y_n}^2 \frac{\sigma_{y_n}^2}{L_n^2}, \tag{5.46}$$

where $\gamma_{y_n}^2$ is the spectral flatness measure for channel no. n, and ϵ_{d_n} is the quantizer performance factor for the difference signal. For the further development it suffices to note that we can find the new gain expression from Equation 5.31 by replacing $\sigma_{y_n}^2$ by $\gamma_{y_n}^2 \sigma_{y_n}^2$, and ϵ_{y_n} by ϵ_{d_n}:

$$G_{SBC-DPCM} = \frac{1}{N} \frac{\epsilon_x}{\epsilon_{d_g}} \left[\prod_{n=0}^{N-1} \gamma_{y_n}^2 \, 2 \int_{\frac{\pi n}{N}}^{\frac{\pi(n+1)}{N}} \frac{S_{xx}(e^{j\omega}) \, d\omega}{\sigma_x^2 \, 2\pi} \right]^{-\frac{1}{N}}. \tag{5.47}$$

ϵ_{d_g} is the geometric mean of the quantizer performance factors of the prediction error signals.

It then follows that the bit allocation formula using optimal DPCM quantization of each band (i.e., with an adapted infinite order predictor) is found by making the same substitutions in Equation 5.19 as above,

$$b_n = \log_2(L_n) = \frac{B}{N} + \frac{1}{2} \log_2\left(\frac{\epsilon_{d_n}}{\epsilon_{d_g}} \frac{\gamma_{y_n} \sigma_{y_n}^2}{\gamma_{y_g} \sigma_{y_g}^2}\right), \qquad n = 0, 1, \ldots, N-1. \quad (5.48)$$

γ_{y_g} is the geometric mean of the channel spectral flatness measures. As $\gamma_{y_n}^2 \sigma_{y_n}^2$ is the average power of the prediction error in channel no. n, the bit allocation can be based on estimates of the prediction error in each channel from blocks of data.

We would like to verify that Equation 5.47 always leads to the optimal prediction gain of a DPCM system. The psd in channel n before downsampling by N is given as

$$S_{y_n y_n}(e^{j\omega}) = S_{xx}(e^{j\omega})|H_n(e^{j\omega})|^2 = \begin{cases} N S_{xx}(e^{j\omega}) \text{ for } \omega \in \pm[\frac{\pi n}{N}, \frac{\pi(n+1)}{N}] \\ 0 \text{ otherwise.} \end{cases}$$

$$(5.49)$$

As always, we assume that the channels are arranged in such a way that we can use integer-band decimation by N to create a baseband representations of the subband signals. Because of the ideal filters used here, there will be no aliasing components, and the decimated spectrum in channel n is thus

$$S_{y_n y_n}(e^{j\omega'}) = N S_{xx}(e^{j(\omega' \pm \pi n)/N}) \text{ for } \omega' \in [0, \pm\pi]. \quad (5.50)$$

The upper signs correspond, and so do the lower signs.

From Equation 5.50 the spectral flatness measure in channel no. n is found as

$$\gamma_{y_n}^2 = \frac{\exp\left\{2 \int_0^\pi \ln\left[N S_{xx}(e^{j(\omega' + \pi n)/N})\right] \frac{d\omega'}{2\pi}\right\}}{2 \int_0^\pi N S_{xx}(e^{j(\omega' + \pi n)/N}) \frac{d\omega'}{2\pi}}. \quad (5.51)$$

Making the change of variable $\omega = (\omega' + \pi n)/N$, we obtain the following expression for the spectral flatness measure:

$$\gamma_{y_n}^2 = \frac{\exp\left\{2N \int_{\frac{\pi n}{N}}^{\frac{\pi(n+1)}{N}} \ln\left[S_{xx}(e^{j\omega})\right] \frac{d\omega}{2\pi}\right\}}{2N \int_{\frac{\pi n}{N}}^{\frac{\pi(n+1)}{N}} S_{xx}(e^{j\omega}) \frac{d\omega}{2\pi}}. \quad (5.52)$$

Inserting this expression into 5.47 and cancelling out terms, we obtain

$$G_{SBC-DPCM} = \frac{\epsilon_x}{\epsilon_{d_g}} \sigma_x^2 \left[\prod_{n=0}^{N-1} \exp\left\{2N \int_{\frac{\pi n}{N}}^{\frac{\pi(n+1)}{N}} \ln S_{xx}(e^{j\omega}) \frac{d\omega}{2\pi}\right\}\right]^{-\frac{1}{N}}$$

$$= \frac{\epsilon_x}{\epsilon_{d_g}} \sigma_x^2 \exp\left[-\sum_{n=0}^{N-1} 2 \int_{\frac{\pi n}{N}}^{\frac{\pi(n+1)}{N}} \ln S_{xx}(e^{j\omega}) \frac{d\omega}{2\pi} \right]$$

$$= \frac{\epsilon_x}{\epsilon_{d_g}} \sigma_x^2 \exp\left[-\int_{-\pi}^{\pi} \ln S_{xx}(e^{j\omega}) \frac{d\omega}{2\pi} \right] = \frac{\epsilon_x}{\epsilon_{d_g}} \gamma_x^{-2}. \qquad (5.53)$$

That is, the maximum gain is inversely proportional to the spectral flatness measure of the input signal independent of the number of channels. This is exactly the prediction gain limit for any coding scheme for Gaussian signals. (That is, the gain from exploiting second order signal statistics). For non-Gaussian input signals, we have to include the ration between the quantizer performance factors in the comparison.

It may puzzle the reader that we are not strongly advocating the above scheme in practical coders. The actual merits of subband coders based on unitary filters (that are supposed to approximate ideal brick wall filters) and DPCM coders in each band do not come close to the theoretical expression. Why is this?

A first question is, why would one use subband signal decomposition prior to DPCM coding at all? The theoretical limit can be reached without the subband part. The answer to this question seems to be that the adaptivity obtained in subband coders through bit allocation is better suited to maximize practical coding gains than adaptive predictors. The bit allocation is also flexible in the sense that it can provide different bit rates to different image parts. To construct adaptive predictors within the subbands seems to be equally difficult. On the other hand, DPCM schemes perform fairly well for speech. Speech is, however, a source which is quite well modeled by AR- or ARMA processes. The same cannot be said for images. At least there is no physical reason to claim such a model for images, as there is no source that generates images. What we have to cope with in image modeling is a signal which is highly nonstationary, and some image parts are equally well modeled by deterministic functions. In the discussion in Chapter 1 we postulated that deterministic structures, such as man-made objects, should be rendered more exactly than stochastic image parts, even though the stochastic parts may contain much more information from a theoretical point of view.

The only popular DPCM application in subband coders seems to be for quantization of the lowpass-lowpass band. This is due to the strong remaining correlations in this band after applying standard filter banks. This is discussed further in Chapter 6.

The comparison between coding methods in terms of series expansions from Chapter 1 may give some further insight into this difficult problem. But subjective aspects must be taken into consideration. We believe that experience only can tell the final truth about the relative performance of coders. Up to this moment there seems to be consensus about the superiority of frequency domain coders over predictive coders.

5.4 Lapped Orthogonal Transforms (LOTs)

The presence of blocking effects in low bit rate coding of images using block transforms inspired several researchers to modify the nonoverlapping transform coding concept by designing *lapped orthogonal transforms*. As the name indicates, this is nothing but a filter bank utilizing relatively short filter responses.

In contrast to traditional filter bank design, a LOT is usually designed to have large coding gain. Cassereau et al. [76] obtained several LOTs using a Lagrange-type iterative search algorithm. The AR(1), $\rho = 0.95$ coding gain was optimized under the restrictions of a unitary filter bank and limited lengths of the filter responses. Malvar and Staelin [52] developed this concept further with a more direct approach to LOT design. In the following we give a brief review of this method.

5.4.1 LOT Design by Eigenvalue Formulation

As in Chapter 2, we assume that the analysis filter responses are given as the columns of the $\mathbf{H}$ matrix. Furthermore, assume that the maximum filter length is restricted, $L \leq 2N$, where N is the number of channels in a uniform filter bank.

For the filter bank to be orthogonal, the columns of $\mathbf{H}$ must be orthogonal:

$$\mathbf{H}^T \mathbf{H} = \mathbf{I}_N, \tag{5.54}$$

and the overlapping basis functions must also be orthogonal:

$$\mathbf{H}^T \mathbf{S} \mathbf{H} = \mathbf{H}^T \mathbf{S}^T \mathbf{H} = \mathbf{0}. \tag{5.55}$$

In the equations above, $\mathbf{I}_N$ is the $N \times N$ identity matrix and $\mathbf{S}$ is a shift matrix defined by:

$$\mathbf{S} = \begin{bmatrix} \mathbf{0} & \mathbf{I}_{L-N} \\ \mathbf{0} & \mathbf{0} \end{bmatrix}. \tag{5.56}$$

Equations 5.54, 5.55, and 5.56 define a unitary filter bank system for the case when $L \leq 2N$. For longer filter responses, the overlap problem becomes more complex because also more distant blocks will influence the current block. Restricting the length to $L \leq 2N$ means that we only have to consider the two neighboring blocks.

The design problem is how to design filter responses fulfilling Equations 5.54 and 5.55 while maximizing the coding gain. In [52] it was observed that given a permissible $\mathbf{H}$ – i.e. a matrix fulfilling Equations 5.54 and 5.55 – it can be modified by

$$\mathbf{H}_0 = \mathbf{H}\mathbf{U}, \tag{5.57}$$

where $\mathbf{U}$ is an arbitrary orthogonal matrix. It follows that Equations 5.54 and 5.55 are fulfilled also for the new filter bank:

$$\begin{align} \mathbf{H}_0^T \mathbf{H}_0 &= \mathbf{U}^T \mathbf{H}^T \mathbf{H}\mathbf{U} = \mathbf{I}_N \tag{5.58} \\ \mathbf{H}_0^T \mathbf{S}\mathbf{H}_0 &= \mathbf{U}^T \mathbf{H}^T \mathbf{S}\mathbf{H}\mathbf{U} = \mathbf{0} \tag{5.59} \\ \mathbf{H}_0^T \mathbf{S}^T \mathbf{H}_0 &= \mathbf{U}^T \mathbf{H}^T \mathbf{S}^T \mathbf{H}\mathbf{U} = \mathbf{0}. \tag{5.60} \end{align}$$

For the modified analysis filter bank $\mathbf{H}_0$, the variances $\sigma_{y_n}^2$ in the coding gain equation for G_{SBC} are given from Equation 5.7, here rewritten as the diagonal entries of the matrix

$$\mathbf{R}_0 = \mathbf{H}_0^T \mathbf{R}_{xx} \mathbf{H}_0, \tag{5.61}$$

where $\mathbf{R}_{xx}$ is the autocorrelation matrix of the input signal.

Substituting Equation 5.57 into Equation 5.61 gives

$$\mathbf{R}_0 = \mathbf{U}^T \mathbf{H}^T \mathbf{R}_{xx} \mathbf{H}\mathbf{U}. \tag{5.62}$$

Since $\mathbf{H}$ and $\mathbf{R}_{xx}$ are fixed, it follows [10] that G_{SBC} is maximized when $\mathbf{R}_0$ is diagonal – equivalently that the rows of $\mathbf{U}$ are the eigenvectors of $\mathbf{H}^T \mathbf{R}_{xx} \mathbf{H}$. Thus, the optimization problem has been reduced to solving the eigenvector problem for a given $\mathbf{H}$. It should be emphasized that optimality is tied to the initial choice of $\mathbf{H}$, i.e. a different choice of $\mathbf{H}$ results in a different solution.

In [52] Malvar and Staelin designed LOTs using an initial filter bank **H** constructed by modifying the basis functions of the DCT. For this choice, optimal LOTs were constructed. Chapter 7 presents coding results using the 16-tap, 8-channel optimal LOT designed in this way, whereas in Chapter 8 we will compare the designed nonunitary filter banks with the same LOT.

Further research in lapped transforms has resulted in a substantially enlarged family – the so-called *modulated* (MLT) and *extended* (ELT) lapped transforms. The MLT is also restricted to filter lengths of $2N$ and the filter responses are not linear phase as in the generic LOT case. The MLT is a cosine-modulated filter bank, and therefore offers low filtering complexity. The ELT is a generalization of the MLT, allowing arbitrarily long filter responses. For further information on LOTs, MLTs, and ELTs, refer to Malvar's book [75].

5.5 Summary

This chapter has covered the important topic of gain optimization in subband- and transform coders. The maximum gain over PCM can be obtained by using a combination of some bit efficient quantization scheme, where the available bits are distributed to the components according to some importance criterion, and a filter bank adapted to the signal characteristics. The optimal filter bank should perform ideal band splitting between adjacent frequency regions to make the channels uncorrelated, and half-whitening within each band to optimize the coding gain. This last criterion is based on the use of scalar quantization after the subband decomposition. The quantization noise in the reconstructed signal will therefore be colored within each subband. An alternative is to apply closed loop DPCM coding within each band. In that case, the optimal filter bank consists of ideal brick-wall filters. The gain is better in that case provided the predictors are optimal. The optimality of the predictors may, however, be a problem. Even if the input process is an AR(1) source, the subbands will, in theory, be AR-sources of infinite order due to the ideal band limitation.

It is important to observe that the subband coder performs significantly better than a transform coder with the same number of bands, especially when the number of bands is small. Whether this is a realistic comparison is debatable. The real comparison should be the subjectively obtainable quality in the two systems at a given bit rate irrespective of complexity.

Implementation complexity is a second issue if the complexity becomes prohibitive. Then a natural comparison would be between a transform coder and the best implementable subband coder.

The coding gain, by itself, does not give correct indications about the performance of a coding system. There are other issues which may be as important. The gain advantage of a subband coder will play an important role in the optimization, but besides from gain the noise type in different coders will be important. Chapter 8 will give a more detailed treatment of these issues.

Chapter 6

Classical Subband Coding

In this chapter we carry on the discussion from Chapter 4 on converting the subband signals or, in the language of transform coding, the transform coefficients into a sequence of bits. We assume that a suitable filter bank/transform has been chosen and that the image has been decomposed using the basis vectors associated with this transform.

Our goal is, as always, to use as few bits as possible to achieve a desired quality of the reconstructed image. We here explore this issue from a practical point of view. Through experiments on natural images, we seek knowledge of the statistical characteristics of the signal entities to be quantized. Thus, we present experimental data showing the intra- and inter-band correlation properties of various image subbands. The form of the *pdf* of the various subbands is also discussed. These issues are dealt with in Section 6.1.

When the statistical properties for each subband signal have been found, we could simply decide to use a suitable quantizer for each band and process each band independently. There are however something to gain by recognizing the origin of the subband signals: They all represent a frequency band of the same real-world image signal. As such, they can be expected to be related in some way. This may be exploited through vector quantization in which the vectors are formed by combining subband samples in different subbands [77, 67, 68, 69]. Simpler, and more common, methods for quantizing the subbands involve scalar quantizers. Two such practical and classical methods involving bit allocation or quantization with a uniform "dead-zone" quantizer are described in Sections 6.2.2 and 6.2.3, respectively.

In Section 6.2.2 we apply the explicit bit allocation strategy presented

187

in the latter part of Section 4.3.1 to subband images. As the name suggests, the basic idea is to allocate bits to the subband quantizers according to the variance in each band. The subband signals are nonstationary in the sense that data statistics vary within each band. In Section 6.2.2 this fact is exploited using an adaptive version of the bit allocation algorithm. Here the bits are distributed *across* the subbands according to the signal energy of local blocks. Whether the bits are distributed on a full subband basis or to local blocks within the subbands, the explicit bit allocation strategies have the advantage, compared to the method using a "dead-zone" quantizer (which makes use of variable length codewords), that the resulting average bit rate is fixed and predetermined irrespective of the characteristics of the image. Also the bit stream is less sensitive to possible transmission errors than when using the method employing a "dead-zone" quantizer, since the latter makes use of variable length codewords.

Independent of which coding strategy is used for the subbands, the low-low band is treated separately because of its intra-sample correlation. In addition, decoded images are very sensitive to errors in this band and high precision quantization is necessary. Section 6.2.1 shows how to exploit intra-sample correlation through DPCM.

The methods introduced so far are designed for quantization of grey-scale images. The generalization to color images is trivial, and Section 6.3 closes this chapter with some comments on this issue.

6.1 Subband Statistics

From Chapter 4 it is evident that the statistical properties of the subband signals play an important role in the selection of a coding/quantization strategy. This justifies a study of these properties for the image subband signals. With this added knowledge we can more easily evaluate a given quantization scheme in terms of a gain-to-complexity ratio, i.e. decide whether a complex scheme really pays off compared to a simple one. In particular we shall study:

- Intra-band correlations.

- Inter-band correlations.

- Probability distributions.

We will treat these issues from an experimental point of view. As a representative coder structure, we have used an 8×8 uniform subband decomposition using the $f\_2\_2\_06$ IIR prototype filter[1] in a tree structure of two-band building blocks. When referring to a particular subband of this two-dimensional decomposition, we use the notational convention exemplified by $(3, 5)$ for subband no. 3 vertically and subband no. 5 horizontally. Since the numbering scheme starts at 0, the low-low band can alternatively be referred to as subband $(0, 0)$. The essential features of the observations to be made in the following are not strongly dependent on the exact choice of a particular subband decomposition scheme.

6.1.1 Intra-Band Correlations

In the following we estimate the degree of sample to sample correlation for some representative subband signals. To estimate the correlation properties one can use the *sample autocorrelation function*[2] [10, 79],

$$
r_{k,l}(u, v) = \frac{1}{KL} \sum_{m=0}^{K-|u|-1} \sum_{n=0}^{L-|v|-1} x_{k,l}(m, n) x_{k,l}(m + |u|, n + |v|), \quad (6.1)
$$

where K and L are the spatial extents in the vertical and horizontal directions, respectively, of subband no. (k, l), which is denoted by $x_{k,l}(m, n)$[3]. The *normalized* sample autocorrelation, which we use when presenting results, is defined by

$$
\rho_{k,l}(u, v) = \frac{r_{k,l}(u, v)}{r_{k,l}(0, 0)}. \quad (6.2)
$$

---

[1]See Appendix C for filter coefficients. Other filter bank choices give similar results. This includes the $f\_2\_2\_x$ and Johnston filter banks whose coefficients are given in Appendix C.

[2]All subband signals, except the low-low band, have approximately zero mean value. For this band we subtract the mean before computing the correlations. Thus, we could just as well speak about autocovariance. We also point out that although the sample autocorrelation is a biased estimate of the autocorrelation, it will be sufficient for our purposes since we only concern ourselves with lags considerably smaller than K and L [78].

[3]k and l denote vertical and horizontal subband indices whereas m and n denote spatial coordinates within a single subband.

Low-Low Band

In Figure 6.1 we show 2-D sample autocorrelation functions for the low-low band of the images "Lenna" and "Boat". For both images we observe substantial correlation between neighboring pixels both vertically and horizontally. Also, in Figure 6.1 we have shown both vertical and horizontal cross sections through the peaks of the 2-D autocorrelation functions. These correlation functions exhibit the same characteristics as natural images in general. This should not come as a surprise, since the low-low band is just a decimated (lowpass filtered and subsampled) version of the original image.

Low-High and High-Low Bands

By a low-high band we shall mean a subband that is lowpass in the vertical direction and non-lowpass in the horizontal direction. A high-low band is non-lowpass vertically and lowpass horizontally. In Figures 6.2 a) and b) we show the vertical and horizontal cross sections of the autocorrelation for subband no. $(0, 3)$ for "Lenna" and "Boat". It is worth noting that there is still some vertical sample to sample correlation present, whereas the sample to sample correlation in the horizontal direction has, for practical purposes, vanished. Similar observations were made for other bands in this category.

High-High Bands

Subbands that are not lowpass in either direction are referred to as high-high bands. As a representative we shall select band no. $(3, 3)$. The correlation functions for this band are shown in Figures 6.2 c) and d). We observe that there does not seem to be any correlation in neither the horizontal nor the vertical direction.

Summarizing the above, we can make some general statements:

- The low-low band has substantial sample to sample correlation in both the horizontal and vertical direction.

- The high-low and low-high bands have some remaining correlation in one direction (the "low" direction), and none in the other.

- The high-high bands have negligible sample to sample correlation.

Based on this we may conclude that it is appropriate to code the low-low band using some method exploiting the high sample to sample correlation.

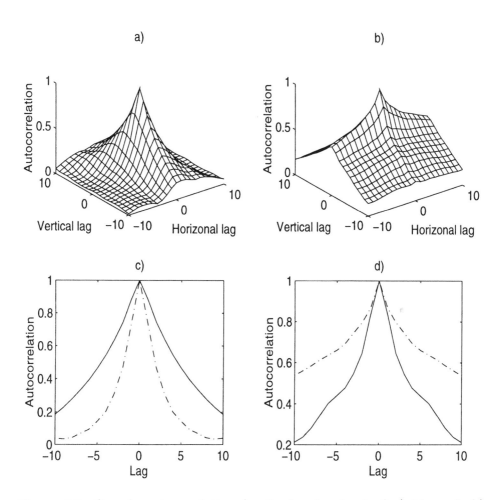

Figure 6.1: *Sample autocorrelation for the low-low band of a) "Lenna", b) "Boat". Horizontal and vertical cross sections of the sample autocorrelation for the low-low band of c) "Lenna" , d) "Boat". Dotted line for horizontal correlation.*

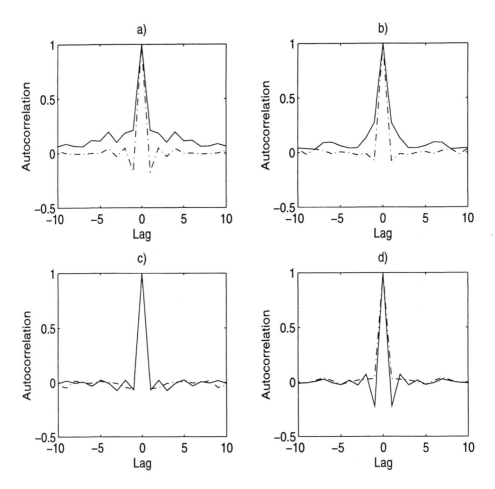

Figure 6.2: *Horizontal and vertical cross sections of the sample autocorre-lation for subband no. $(0,3)$ of a) "Lenna" , b) "Boat". Horizontal and vertical cross sections of the sample autocorrelation for subband no. $(3,3)$ of c) "Lenna" , d) "Boat". Dotted line for horizontal correlation.*

The use of a DPCM coder for this band is quite common. For the other bands straight forward scalar quantization is appropriate, *if* the subbands are to be coded separately. The remaining correlations in one direction for low-high and high-low bands are not high enough to warrant a DPCM scheme.

6.1.2 Inter-Band Correlations

If we assume that our filter banks have an ideal frequency response – i.e. the filters' transition regions are infinitely steep and there is no overlap between the frequency bands, the cross spectral densities between the subbands are equal to zero. This implies that the crosscorrelations between the subbands is equal to zero, see Section 5.1.2. Although, our filter banks are not ideal, we do expect a crosscorrelation that is close to zero. We substantiate this through some experiments.

As a measure of the dependencies between different subbands, we shall use the sample crosscorrelation. For subbands $x_{k,l}(m,n)$ and $x_{\xi,\nu}(m,n)$ this function is defined by

$$r_{kl,\xi\nu}(u,v) = \frac{1}{KL} \sum_{m=0}^{K-|u|-1} \sum_{n=0}^{L-|v|-1} x_{k,l}(m,n) x_{\xi,\nu}(m+|u|,n+|v|), \quad (6.3)$$

where we have tacitly assumed that the two subband signals have the same spatial extent. Again, we shall find it more convenient to operate with normalized functions, which in the present case leads to the following definition:

$$\rho_{kl,\xi\nu}(u,v) = \frac{r_{kl,\xi\nu}(u,v)}{\sqrt{r_{k,l}(0,0)r_{\xi,\nu}(0,0)}}. \quad (6.4)$$

Note that values close to zero imply small crosscorrelation, whereas values close to 1 or −1 imply high positive or negative correlations, respectively.

We have investigated the crosscorrelations between various subbands for both IIR and FIR[4] tree-structured uniform filter banks with 8×8 subbands. As representative results we present the sample crosscorrelations between bands (0,1) and (0,2), (1,1) and (1,2), and between bands (0,1) and (1,2). We have summarized the experimental results, which were obtained from the subbands of the image "Lenna", in Table 6.1. Based on these experiments

---

[4] Johnston type prototype filters [17].

Subbands	$\rho(0,0)$	ρ_{max}	ρ_{min}
(0,1) - (0,2)	-0.11	0.04	-0.12
(1,1) - (1,2)	0.02	0.07	-0.08
(0,1) - (1,2)	0.04	0.06	-0.06

Table 6.1: *Cross correlations between selected subbands for an* 8×8 *subband decomposition of the image "Lenna".*

we conclude that the crosscorrelations between subband signals in our filter banks are either negligible or small. Thus, separate coding/quantization of each subband is an appropriate approach.

6.1.3 Probability Distributions

From Chapter 4 it is evident that if the various subbands are to be quantized with separate quantizers, and if the number of representative levels is to be determined by some bit allocation strategy, the design of pdf optimized quantizers is appropriate. Thus, some knowledge of the form of the pdf of the various subbands is necessary. Several authors [80, 77, 51] have investigated this issue experimentally by examining histograms of various subbands using different filter banks and subband decompositions of many different images. It is concluded that most subbands can be modeled very well by either a *Laplacian* or a *gamma* pdf. The Laplacian pdf is appropriate for the very high frequency bands, whereas the gamma pdf is most appropriate for the lower frequency bands. As shown in Chapter 4, this "peaky" behavior of the subband pdfs is due to the nonstationary nature of the signal.

Given the high sample to sample correlation of the low-low band, it makes sense to apply a DPCM coder to this band. The pdf of the prediction error, for several predictor configurations, was found to be modeled well by a Laplacian pdf [80, 77, 51]. Thus a quantizer optimized for this pdf will be a good choice.

6.2 Scalar Image Subband Quantization

A large number of subband still image coders have been proposed in the literature. Many employ some sort of scalar quantization of the subbands, whereas others employ vector quantization. In [69] Westerink et al. propose a vector quantization strategy that explicitly makes use of any possible inter-band statistical dependency, as well as sphere packing. They do not, however, claim any performance gains relative to an approach in which a *sophisticated strategy* based on scalar quantization is employed. Several other vector quantization schemes have been proposed lately – see [81] for a short review – and compared to subband coders based on scalar quantization. These comparisons do not always provide a fair comparison of the situation in that the scalar approaches in some of the comparisons are rather simplistic. Under all circumstances, scalar quantization of image subbands, when a relatively large number of subbands is employed, is of great practical significance.

6.2.1 Coding of the Lowpass Band

Due to the high correlation remaining in the low-low subband, scalar quantization of the signal would lead to inferior performance. In principle, we could employ any coding technique suitable for coding of strongly correlated signals – such as the DCT, DPCM or vector quantization. The DCT has, for example, been used by Ansari [82] in coding this band in a subband coding scheme for high definition TV (HDTV) in which the input image was split into 2×2 subbands.

The main issues to be considered in designing a DPCM coder for the low-low band are:

- The predictor configuration, i.e. how many samples to base the prediction on, and their location relative to the location of the sample to predict.

- Should we employ a fixed predictor or is a predictor adapted to the signal to be coded preferable?

- How many bits per pixel should be spent in the lowpass band?

- What type of quantizer should be selected?

All these issues were dealt with in [51]. A review of the conclusions of that
work will be presented below.

In the design of predictors for pure DPCM image coders it has been
observed [83] that, for common images, the decrease in the prediction error
signal is substantial for prediction orders up to 3. Further increases in the
order of the predictor yield only marginal decreases in the prediction error.
Since the lowpass band has properties similar to the original image, it is
argued in [51] that a third order predictor is a good choice. Following [51], a
quantizer optimized for the Laplacian probability density function was used
for quantizing the prediction error signal. The use of an optimal predictor,
calculated for each image to be coded, was also explored. Although the
performance of the coder increased somewhat, it was found that the gain
was so small that the effort involved in computing the optimal predictor was
not justified. Therefore, the fixed predictor of [84] formed according to

$$\hat{x}_{0,0}(m,n) \quad = \quad 0.5x_{0,0}(m, n-1) + 0.25x_{0,0}(m-1, n) \qquad (6.5)$$
$$+0.25x_{0,0}(m-1, n+1),$$

was found to represent a good compromise between performance and complexity. Based on substantial experimental evidence, it was found that about
5 bits per pixel for the lowpass band was the optimal choice for target rates
below 0.5 bits/pixel range[5]. This means that the low-low band accounts for
about 0.08 bits per pixel in the total bit budget. At higher rates, it pays
off to spend 6 bits/pixel in the lowpass band. It was found that further
increase in the number of bits spent on this band does not pay off, even at
substantially higher total bit rates.

6.2.2 Coding of the Higher Bands I: Explicit Bit Allocation

Previously, it was concluded that scalar quantization is suitable for the non-
lowpass bands. In subband coding of both speech and image signals, the
traditional approach has been to allocate quantizer bits to whole subbands
in such a way as to keep the mean square error between the decoded image
and the original image at a minimum. The theoretical aspects of this is dealt
with in Section 4.3.1. In that section it is also pointed out that due to some
practical problems associated with the theoretical bit allocation formulas,
an iterative algorithm for distributing the bits might be a viable solution.

---

[5]The images coded were of size 512×512.

Westerink [77, 69] makes use of an explicit bit allocation algorithm that distributes bits on a full subband basis. The algorithm differs somewhat from that presented towards the end of Section 4.3.1 in that it is based on considering a large collection of permissible quantizers for the subbands. For a given assignment of quantizers to the subbands a distortion function is evaluated. Conceptually, this is done for all possible assignments of permissible quantizers to the subbands. For all the quantizer assignments, the computed distortion vs. the bit rate are entered into a scatter plot. By finding the *convex hull* of this scatter plot, a rate-distortion curve, $R(D)$, is found. This curve gives the minimum distortion for a given bit rate, R.

Block-Wise Adaptive Bit Allocation

The strategies described above use the same type of quantizer for all the samples of a given subband. Thus, the variation of the image activity within a single subband is not fully exploited.

Recognizing the nonstationary character of the subbands of real images it may be advantageous to make the quantization adaptive to local characteristics of each subband. This can be done by dividing each subband image into blocks of size $k \times k$. Rather than having the complete subbands compete for bits, each $k \times k$ block is allowed to compete for bits.

The bits are distributed as outlined in Section 4.3.1:

1. An estimate of the standard deviation, σ_m, of each $k \times k$ block in all the subband images is computed.

2. The block, say block no. l, with the highest estimated standard deviation is assigned one bit.

3. The estimate of the standard deviation (σ_l) associated with the block having just been allocated a bit is replaced by $\sigma_l/2$.

4. If all available bits have been allocated, then stop. Otherwise, continue at step 2 above.

Upon completion of this procedure, a bit allocation table for all $k \times k$ blocks of the subband partitioned image has been compiled. At this point all blocks – irrespective of which subband they belong to, having the same number of bits allocated, are quantized employing the same quantizer. Blocks with a given bit assignment, for example p bits, are said to belong to *class*

p. To prevent variance mismatch in the quantization process, the collection of blocks within a class is normalized to unity variance prior to being put through the quantizer. Each class variance, or scaling factor, is transmitted to the decoder as side information. Also, for each block we must indicate the class to which it belongs.

For low bit rates (~ 1.0 bits and below), it was reported in [51] that 6 different block classes, corresponding to quantizers with 0 to 5 bits per pixel, is a suitable choice. Thus, we need to transmit 5 standard deviation values and, for each block, a class number between 0 and 5. This corresponds to sending the bit allocation table. The contribution to the total bit rate of the 5 standard deviation values is negligible. The contribution of the class numbers depends on the choice of k. 4×4 blocks represent a good compromise between a high degree of local adaptivity and the requirements for bits to be spent as side information. With this block size and for target bit rates of around 0.5 bits per pixel, about 0.05 bits per pixel have to be spent for the purpose of indicating block classes [51].

In Figure 6.3 we have, for illustrative purposes, shown a bit allocation map for "Lenna" coded at 0.5 bits/pixel[6]. The low-low band is excluded from the map since it does not compete for bits through the bit allocation algorithm. Also observe that white signifies that 5 bits per pixel have been assigned for a 4×4 block, whereas black indicates that no bits have been assigned. Gray-levels indicate bit assignments in between those extremes.

Quantizer Selection

In the scheme just described the source signal to be quantized by a particular quantizer is made up of signal samples taken from several different subbands. Therefore, we need to know the histograms of the source samples belonging to a *single class*. In [51] it was found that these histograms, for several images and a target bit rate around 0.5 bits/pixel, were very closely approximated by a Laplacian pdf. Thus, quantizers optimized for this pdf were used in the coder.

---

[6]In this example a 28 band split, as shown in Figure 9.24 of Chapter 9, was used.

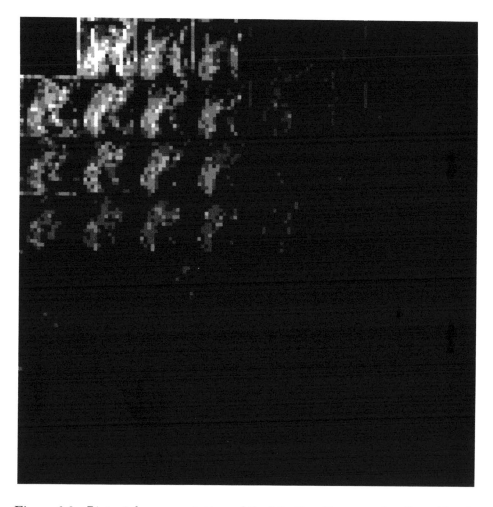

Figure 6.3: *Pictorial representation of the bit allocation map for the subbands of "Lenna". The target bit rate is 0.5 bits per pixel. White signifies 5 bits per pixel at the location of the corresponding block, whereas black indicates that no bits have been assigned. We have not shown the allocation for the low-low band, since this is assumed fixed.*

6.2.3   Coding of the Higher Bands II: Coding using "Dead Zone" Quantization and Subband Scanning

The strategy based on explicit bit allocation, as described in the previous section, is not particularly simple from a computational point of view. In connection with low complexity filter banks such as IIR or short-kernel FIR filter banks, it would be attractive to apply these in a complete coder structure characterized by low *overall* computational complexity. For such purposes the elements from the strategy for coding of the transform coefficients in the scene-adaptive DCT based still image coder of Chen and Pratt [85] are employed [51].

In [85] the image to be coded is split into square blocks that are subsequently transformed by a 2-D DCT of the same size as the block. As we have pointed out before, it is possible to view the DCT as a particular instantiation of a critically sampled filter bank. Thus, the collection of DC coefficients may be interpreted as the low-low band, and the collection of all transform coefficients with a given index may be interpreted as the signal samples of a particular subband. Consequently, an 8×8 2-D DCT is nothing but a parallel separable filter bank with 64 uniform subbands. In the coding scheme of Chen and Pratt [85] the DCT coefficients are scanned in a zig-zag pattern before being quantized, run-length and entropy coded. Interpreting the DCT as a filter bank, the scanning [85] can be explained as follows: For each spatial coordinate (m, n) the pixel value in subband no. (k, l) can be written as $x_{k,l}(m, n)$. This can be viewed as a function of (k, l). For each value of m and n we create a vector of 63 subband samples[7] – one from each subband, in sequence of increasing spatial frequency as illustrated in Figure 6.4. There is, of course, no reason why this scanning strategy should be restricted to "subband coders" using the rows of the DCT matrix as filter bank basis functions. In fact the subbands of any uniform subband decomposition can be scanned in the same fashion.

In the resulting 1-D *scan string*, the low frequency subband samples appear first and the higher frequency subband samples follow as one moves towards the end. Since subband samples of low amplitude are of small perceptual importance, the scan strings are thresholded. That is, those subband samples whose magnitude are below the threshold are set to zero. The coefficients remaining after this operation are quantized by a uniform quantizer.

---

[7]Remember that the low-low band is left out, – this is treated separately with a DPCM coder as explained previously.

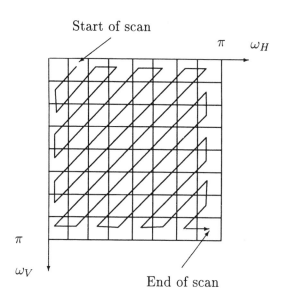

Figure 6.4: *Zig-zag scan of the subbands for a uniform filter bank. An 8 × 8 filter bank with real subbands is assumed. The resulting scan string contains one sample from each subband except the lowpass band.*

The quantizer characteristic for such a quantizer is shown in Figure 4.2. Typically, the quantized scan string will have nonzero values for low string indices, whereas the proportion of zero valued samples tends to increase as one moves towards the end of the string. In [85] it is pointed out that this is a consequence of the scan order. This increasing proportion of zero samples as the string index increases facilitates its efficient representation through *run-length coding* [83]. Finally the run-length codewords are further coded using a variable length entropy code (VLC). The VLC employed is very similar to the one applied in the H.261[86] recommendation for coding of video telephony image sequences. For an explanation of how this works, see Section 9.3.1 of Chapter 9. In fact, that section describes a video coder that is based on the same concepts as the present still image coder.

6.3   Extension to Color Images

Digitized color images are commonly quantized in an RGB format – that is, the red, green, and blue components are represented separately, typically by 8 bits per component pixel. It is well known that the three components are highly correlated [87, 88]. Separate coding of these components would therefore be wasteful. In [89] Kim et al. attempted at creating another set of three uncorrelated color components by means of a Karhunen-Loève transform. They found, however, that there was no gain associated with coding these new color components rather than the luminance-chrominance components, YIQ or YUV, as defined in the NTSC/PAL TV standards. In this terminology Y denotes the luminance component and I/U and Q/V denote the two chrominance components. There are no advantages associated with the YIQ or YUV relative to the other in a coding context.The transformation between the RGB and YIQ systems is given by

$$
\begin{bmatrix} Y \\ I \\ Q \end{bmatrix} = \begin{bmatrix} 0.299 & 0.587 & 0.114 \\ 0.597 & -0.277 & -0.321 \\ 0.213 & -0.523 & 0.309 \end{bmatrix} \begin{bmatrix} R \\ G \\ B \end{bmatrix}
\tag{6.6}
$$

The inverse of this transformation brings one from the YIQ to the RGB domain.

It is well known [87] that the Y component is essentially a monochrome image whereas the I and Q components are images of very low variance. Thus, the the perceptual importance of the I and Q components is much less than that of the Y component. This fact has been exploited by several researchers through filtering and subsampling of the chrominance prior to coding. After decoding, the chrominance components are again brought back to their original size by interpolation. The degree to which the chrominance is subsampled has a significant impact on the total bit rate that needs be spent on the color information. For low bit rate applications subsampling ratios between 4 and 8, both vertically and horizontally, were successfully applied [89, 51].

In [51] each chrominance component was passed through a separate subband coder. For these inputs, all the subbands, except the lowpass band, were set to zero. The lowpass band was coded using DPCM at 3 — 5 bits per pixel. The Y component was treated in exactly the same fashion as a monochrome image. A consequence of this approach is that the proportion of the total bit rate to be expended on each of the Y, I and Q components has to be determined a priori. This is in conformance with the choice of Kim et al. [90, 89], but different from the approach of Westerink [77] in which the various components compete for bits through a bit allocation strategy.

6.4 Summary

In devising strategies for the bit efficient representation of the subband images one needs to know their properties. In this chapter we have seen that:

- The subband signals, except the lowpass band, possess low sample to sample correlations. This implies that PCM is a suitable coding approach for the non-lowpass subbands.

- The faithful representation of the lowpass band is of utmost perceptual importance. This observation combined with this band's high sample to sample correlation led to a DPCM scheme in which a high number of bits (5–6 bits per pixel) is used.

- Different subband signals possess low crosscorrelation. Thus, we can expect good results by treating each subband separately in the coding process.

- The histograms of the signals to be quantized can be modeled by well known probability distributions. This is useful in the selection of optimal quantizer characteristics.

These observations formed the basis for presenting some approaches for coding of the subbands using scalar quantizers. First, we explored coders based on explicit bit allocation. As an alternative to this approach we described in Section 6.2.3 the computationally very attractive coding scheme, involving a "dead-zone" quantizer, for the representation of the subbands. This scheme combined with computationally efficient filter banks results in complete coder structures characterized by extremely low *overall* computational complexity. Finally, the schemes originally introduced for the coding of monochrome images were extended to color images.

Chapter 7

Coding Examples and Comparisons

In Chapter 5 we have focused on theoretical measures for the expected performance of various filter banks and transforms employed in subband coders. In this chapter we present examples of decoded images when using subband coders with different choices of filter banks. The aim is to identify the types of visual degradations associated with particular choices of filter banks.

As representative images in the presentation of decoded images we have selected three very different test images, all of which are reproduced in Appendix D:

1. *"Lenna":* A low complexity image with textured as well as very smooth areas.

2. *"Boat":* A quite detailed image with sharply defined masts and rope against a smooth background.

3. *"Kiel":* An image containing both extremes: High detail area in the bottom half of the image and all-smooth skyline in the upper half. Loosely described as a "subband coder's nightmare image", this suggests the problems of representing it as linear combinations of basis functions corresponding to *long* filter unit pulse responses. In fact, this is an ideal image for demonstrating a transform coder: All smooth sky region (not a cloud in sight) gives no problems with blocking effects. The extremely detailed lower half is very well suited for the short unit

pulse responses of a transform coder. Blocking effects pose no problems in this region either, because they are efficiently masked.

An investigation of the visual performance of different image coding methods must necessarily be incomplete as well as subjective. Acknowledging this, we try to highlight pros and cons of some of the possible filter bank choices by showing selected image subregions that demonstrate the main classes of image artifacts, namely blurring, ringing and blocking effects. In addition a comparison of the two signal extension methods discussed in Section 2.4 is presented.

First we present some examples of blurring distortion for one filter bank only, realizing that this distortion does not significantly depend on the exact choice of filter bank. When showing examples of ringing and blocking distortion we employ several filter banks, namely the DCT, a 16-tap LOT designed as shown in Section 5.4.1, a Johnston type FIR filter bank [17] and one of the IIR filter banks of [51]. All these filter banks are treated in Chapters 3 or 5. In all the examples to follow we use a uniform 8×8 band decomposition of the image. When using the Johnston type FIR filter banks and the IIR filter banks, this decomposition is obtained by organizing two-channel building blocks in a tree-structure as explained in Chapter 3.

All experiments were carried out using the same quantization procedure: The lowpass band was DPCM coded with 5 bits per pel using an optimal 3rd order predictor, applying the configuration due to Gharavi and Tabatabai [84], and computed according to the covariance method [79]. A Laplace quantizer was used inside the prediction loop [51]. The quantization algorithm for the upper bands, previously described in Section 6.2.3, is based on thresholding with "dead-zone" twice as large as the quantization step. This is followed by run-length-entropy coding, following the method of [51].

Limiting ourselves to "traditional" subband filter banks in the present chapter, the next chapter will present some approaches to the design of filter banks leading to less disturbing coding artifacts.

We repeat that this demonstration is in no way exhaustive, as it would be possible to include a larger selection of filter banks and combine them with a wider selection of coding strategies.

7.1 Blurring Distortion

In studying the blurring phenomenon, we focus on the choice of threshold value and quantization step, denoted t and Δ in Figure 4.2.

Both Δ and t are obviously related to the desired total bit rate. If t is too small, many high frequency information samples in higher bands will be quantized to values different from zero. This is disadvantageous for the following reasons:

- The total bit rate will be high because the high frequency samples need many bits for representation. The run-length code used here will be inefficient in this situation.

- The quantized high frequency samples may generate unwanted artifacts in visually smooth image areas. When such a region contain weak, high-frequency textures, some subband samples in the higher bands are quantized to nonzero values. Even though the synthesis low pass response reconstruct a smooth underlying surface, some spurious artifacts results from the nonzero higher band samples.

On the other hand, if t is too large, too much image detail will be lost, giving the decoded image a very blurred appearance. Choosing a suitable threshold therefore entails a trade-off: Avoid high bit rate and artifacts in smooth regions, while maintaining fine image details.

To demonstrate the importance of using a suitable value for the threshold level, we code an image using fixed quantization step and a varying threshold level. The value of Δ is chosen to give high image quality – thus highlighting the effect of increased threshold level.

As an example, consider the image "Lenna" coded using the IIR filter bank $f\_2\_2\_06$ given in Appendix C, with $t = \Delta, 2\Delta, 3\Delta$ and 4Δ. The bit rates and corresponding SNRs are given in Table 7.1, and in Figure 7.1 selected image subregions are shown. The LP-LP band was coded with DPCM at 5 bits per pel.

As evident from Table 7.1, Δ is small enough to give a signal-to-noise ratio of 36.1 dB when the dead-zone is twice the size of the quantization step $(t = \Delta)$.

We have picked two image subregions for demonstrating the above mentioned trade-off. For $t = \Delta$ the details of the hat texture are well preserved. As t is increased, more texture detail is lost and the shoulder area becomes

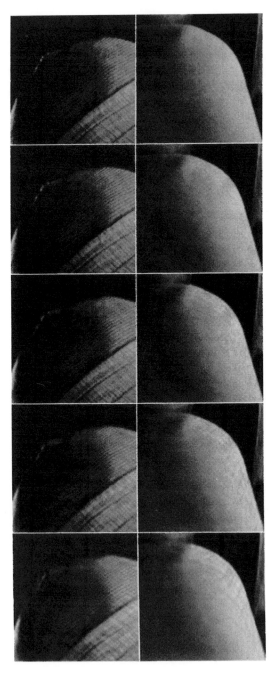

Figure 7.1: *Varying the threshold: Original (top), $t = \Delta$ to $t = 4\Delta$ (bottom).*

Threshold t	Δ	2Δ	3Δ	4Δ
Bit rate	0.81	0.36	0.23	0.16
SNR (dB)	36.1	33.4	31.9	30.8

Table 7.1: *Bit rates and SNRs for coding of "Lenna" with filter bank* f_2_2_06 *at several bit rates.*

slightly smoother. In addition, this filter bank exhibits excessive ringing noise. Note the difference in bit rate between $t = \Delta$ and $t = 2\Delta$. Clearly, the expense in terms of bit rate is not worthwhile.

In general, for low bit rate coding of images, the quantization step will be higher than for this experiment, and as a rule of thumb a threshold level equal to the quantization step is often used.

7.2 Ringing Noise

Ringing noise in subband coders is generated in regions with sharp transitions. This artifact is closely related to the *Gibbs phenomenon* [78] that arises in the convergence analysis of Fourier series. Since ringing noise is masked in textured regions as well as in regions with many image details, we show decoded image subregions that have transitions in the vicinity of smooth areas.

Good examples of the nature of the ringing distortion can be seen by coding the "Boat" image. Figure 7.2 shows the original image excerpt, whereas Figures 7.3 and 7.4 depict the coding results obtained when using $t = \Delta$ in conjunction with several filter banks. The bit rates are in the 0.43 to 0.46 bits per pel range.

From these figures it is evident that the ringing distortion is most prominent in traditional subband coders employing filter banks with long unit pulse responses – in this case the ones based on the *f16b* FIR and the *f_2_2_06* IIR filters[1]. The DCT and the 16-tap LOT exhibit very small amounts of ringing distortion.

---

[1]See Appendix C for filter coefficients.

Figure 7.2: *Original "Boat" image (excerpt).*

7.3   Blocking Distortion

In transform coders blocks of the image to be coded are processed and represented independently. In traditional subband coders the image is processed as one single signal entity. The LOT may be considered as an intermediate case in this context. The independence in processing and representation of adjacent blocks in a transform coding scheme manifests itself as blocking distortion. As this independence is reduced in LOTs and even more so in traditional subband coders (i.e. subband coders employing channel filters with long unit pulse responses) the blocking distortion is reduced and, for the case of traditional subband coders, it has completely vanished.

These statements are illustrated in Figure 7.6 and Figure 7.7 (as well as in Figure 7.3, top). For comparison the excerpt of the original of "Lenna" used in these coding experiments is shown in Figure 7.5.

7.4   Comparing Circular and Mirror Extension

In Figure 7.8 we show two coding examples taken from the image "Lenna", where we compare the circular extension method with the mirror extension

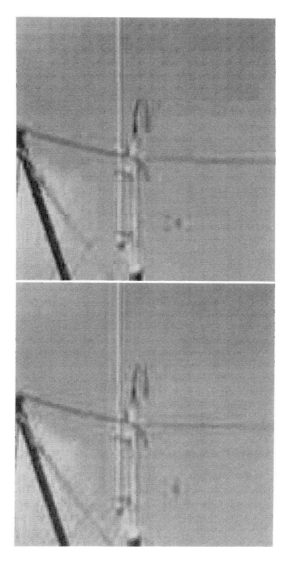

Figure 7.3: *Ringing noise for various filter banks: DCT (top), and LOT (bottom).*

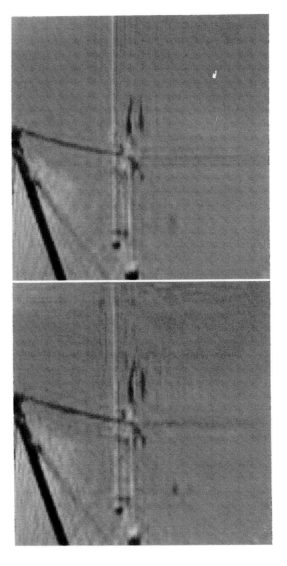

Figure 7.4: *Ringing noise for various filter banks: Johnston FIR filter bank, f16b (top), and IIR filter f_2_2_06 (bottom).*

Figure 7.5: *Original "Lenna" image (excerpt).*

method. The complete image is filtered with the f32d filters due to John-ston [17]. The total bit rate is 0.25 bit per pel. The improved result of the mirror extension method is clearly evident in the left, right, and top edge regions. These are the regions affected by the generation of artificial steps when using the circular extension method. Similar problems affects the bot-tom edge region but this is omitted from the excerpts shown here. Although there is a clear quality difference in this case, we remark that the difference is not very significant for higher bit rates.

Figure 7.6: *Blocking distortion for various filter banks: DCT (top), and LOT (bottom). Bit rate: Approximately 0.4 bits per pel.*

Figure 7.7: *Blocking distortion (or lack thereof) for various filter banks: Johnston FIR filter bank, f16b (top), and IIR filter f_2_2_06 (bottom). Bit rate: Approximately 0.4 bits per pel*

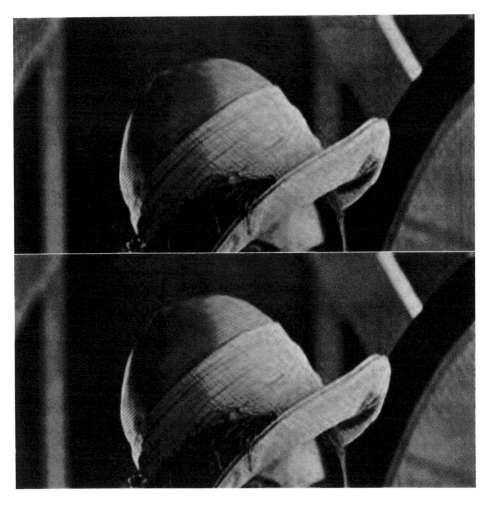

Figure 7.8: *Comparison of signal extension techniques:Circular extension (top), and mirror extension (bottom).*

Chapter 8

Remedies for Coding Artifacts

At high compression ratios, the image transform/subband data are coarsely approximated. As a result of the associated quantization noise, the decoded image will contain distortions. In Chapter 7 we have provided examples of how the reconstructed images are degraded as the total bit rate is decreased, with the most annoying distortions shown in Figure 7.4 and Figure 7.6 for ringing distortion and blocking effects, respectively.

The perceived distortions are a combination of image artifacts and generalized blurring or lack of detail. Artifacts are particularly annoying, and it is therefore important to develop coding algorithms capable of achieving large compression ratios while preserving the natural appearance of the original image. Obviously, some distortion must be expected when coding images at low bit rates. A good coding algorithm should therefore be robust in the sense that no strange artifacts appear in the decoded images when the bit rate is further decreased. One way of achieving this goal is to allow for artifacts in the decoded image, and subsequently apply some kind of image postprocessing. An algorithm presented in [91] uses the original image for extraction of crude information about smooth regions, and transmits this side information along with the quantized subband signals. After normal subband decoding, the side information is used to process the decoded image with lowpass filter operators in the smooth regions. Since ringing noise is effectively masked in textures and close to edges, this method works well at low bit rates. In this chapter we will in particular study how the choice of filter bank affects the noise characteristics. This investigation leads to criteria for optimization and evaluation of a filter bank's performance in a low bit rate context.

Historically, there has been a clear distinction between transform and subband coding. This can be related to the fact that traditional subband coders use long unit pulse responses, giving rise to the notion of frequency band splitting. In Chapter 5 we saw that the performance of a subband coder having ideal frequency band partitioning approaches that of the ideal prediction gain when the number of subbands approaches infinity. In addition, it was shown that using perfect frequency splitting, two adjacent subband signals will be perfectly decorrelated. In practice, however, the typical number of bands used in each image dimension is limited to 8 or at the very most 16, and the frequency selectivity is not perfect. Early filter bank designers were concerned with obtaining good stop band attenuation, little ripple in the pass band, and low aliasing between adjacent bands [17]. Therefore, the resulting filter bank unit pulse responses were long.

The transform coding community had a very different theoretical framework. The optimal transform in terms of theoretical coding gain is the *Karhunen-Loève* transform [87]. Optimality is in the mean square sense. This transform is simply the eigenvector matrix of the signal autocorrelation matrix, and is therefore signal dependent (see Section 5.1.1). For images, a good, signal independent transform is the DCT, which is very close to the Karhunen-Loève transform of a first order autoregressive (AR(1)) process with correlation coefficient approaching unity. Thus, assuming that the AR(1) process is a satisfactory image model, all is well.

This was status quo until 1989 when Cassereau et al. [76] and Malvar and Staelin [52] merged the two disciplines with the advent of the *Lapped Orthogonal Transform* (LOT). The LOT is nothing but an example of a filter bank featuring *perfect reconstruction* as well as having *orthogonal* unit pulse responses. Before this, it was known that the transform coder can be seen as a special case of the subband coder. The new concept in those two papers was to design filter banks with short unit pulse responses and to use the AR(1) coding gain as criterion for performance instead of the channel separation properties previously mentioned.

In this chapter we shall mainly be concerned with the filter bank's influence on the visual appearance of decoded images at low bit rates. Some work has been done in this field, particularly for reducing ringing noise, which is the dominant noise component in traditional SBC [92, 93]. With the introduction of the LOT, this noise type is reduced at the cost of blocking effects.

Following [22], we try to minimize these noise components simultaneously.

Although most of the material in this chapter deals with improvements in filter bank design, another important trend that is discussed is new insight in bit-efficient signal representation. One technique based on the dynamic entropy coder allocation from Section 4.3.3 is demonstrated at the end of this chapter.

8.1 Classification of Image Defects

As in Chapter 7, we will in the following utilize a simple, but useful noise classification, similar to that of [94]. We divide the noise components into 3 categories:

- Blurring.

- Blocking effects.

- Ringing noise.

Blurring results from the cancellation of high frequency image details by skipping the higher bands, and cannot be avoided at low bit rates. This was demonstrated in Figure 7.1. Allocating more bits to the higher bands simply reduces the blurring at the cost of increased noise at other frequencies. Fortunately, moderate blurring is not considered annoying by a human observer since it is a "natural" type of distortion.

Blocking effects and ringing noise are the noise artifacts experienced in transform and traditional subband coders, respectively. When coding the transform coefficients at low bit rates, the quantization errors will lead to the appearance of the blocks that the image is split into. This is a natural consequence of the independent processing of each block. In traditional subband coders the blocking effect is avoided since the basis vectors (unit pulse responses) overlap. Instead, ringing noise is experienced: A quantization noise sample in the subband domain will be filtered through the synthesis filter bank, and the reconstruction noise will be spread over an area given by the unit pulse response length. A plausible unit pulse response length for a traditional filter bank may extend well beyond 100 pels, while for a standard 8×8 block transform coder the unit pulse response length is 8. For this reason, transform coders have negligible ringing noise.

For the new generation of filter banks having shorter unit pulse responses, for example the LOTs [76, 52], the noise artifacts are of the same character as the blocking effects of transform coding, only smoother. Figure 8.1 illustrates how the reconstruction noise varies with the unit pulse response length in an artificial but revealing example. In this experiment, we preserved the lowpass band, whereas the higher bands were canceled out. This rather dubious method of "quantization" was applied for presentation purposes. The original ramp transition signal is reconstructed using long (32-taps [17], 3-stage hierarchical structure, resulting in an effective filter length of 218 taps) and short (16-tap, LOT [52]) unit pulse responses in conjunction with $N = 8$ bands.

The ringing noise is evident for the 218-tap reconstruction in the vicinity of the transition, and results from the over- and undershoots in the lowpass filter response. At some distance from the transition, the reconstructed signal is smooth and similar to the original. In practice, the ringing noise will be well masked by textures and edges while perfectly visible (and indeed annoying) in smooth image areas.

For the 16-tap LOT the reconstruction noise has an appearance similar to blocking effects in a transform coder. In the latter case the reconstructed ramp signal would resemble a simple staircase pattern. Here, each step is not completely flat, but the visual impression of the reconstructed image is slightly similar. The reader is urged to inspect Figure 7.6 for a visual comparison.

From this discussion it is evident that in choosing a unit pulse response length, a trade-off between blocking effects and ringing noise is made. In the following section, we will focus our attention on the blocking effects. As will be shown in Section 8.2, special care when designing the filter bank will be rewarded in terms of an improvement with respect to this problem. A good trade-off between ringing noise and blocking effects can therefore be obtained if we choose a short unit pulse length, and subsequently optimize the filter taps for reduced blocking effects.

8.1.1   A Study of Blocking Effects

Blocking effects are visible in smooth image areas, whereas in complex areas, such as textures, any underlying blocking is effectively masked. We therefore direct our attention to the reconstruction of smooth areas at low bit rates.

In most subband coding schemes the lowpass-lowpass (LP-LP) band is

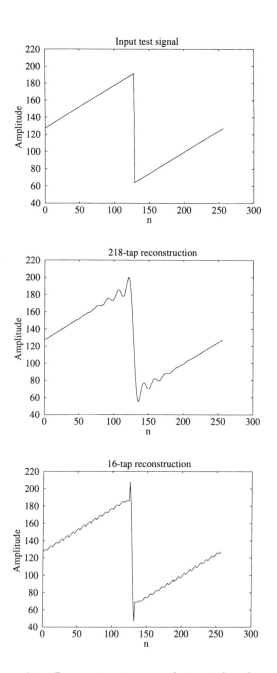

Figure 8.1: *Reconstruction artifacts at low bit rates.*

coded separately at a fixed bit rate. Therefore a restriction on the total amount of bits to be used only affects quantization of the higher bands. If we concentrate on such coding schemes, it is plausible to model the quantization procedure at low bit rates *in smooth image areas* as a basis restriction error [87], where the retained basis vectors are a set of time shifted lowpass responses only. When filtering smooth image regions through the analysis filter bank, most of the signal energy goes into the lowpass band. The signal samples in the higher frequency bands will mostly have small values, and they are therefore cancelled out in the quantization process. Thus, reconstruction of the decoded image region will only depend on the lowpass analysis and synthesis filter responses. The basis restriction procedure is shown in Figure 8.2.

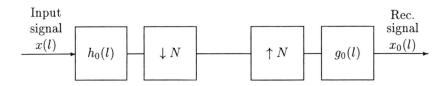

Figure 8.2: *Low bit rate coding modeled as a basis restriction. Only the lowpass channel filters are used.*

If we want to minimize the blocking effects at low bit rates, we should construct the lowpass channel to be a good interpolator of smooth image areas, i.e. the filter bank should be able to reconstruct smooth image areas using lowpass band data only. This approach simplifies the filter bank design procedure compared to that of computing the cumulative effect of all channels with respect to the resulting blocking effects.

Error Analysis

For a one-dimensional input signal $x(l)$ we now want to study the reconstructed signal $x_0(l)$ when only the lowpass samples are transmitted to the decoder. We assume a uniform, N-band, critically decimated filter bank with unit pulse response lengths $L = pN$, where p is a positive integer. Using the input-output equations of the analysis and synthesis filter banks,

Equations 2.33 and 2.35, we can, after some manipulations, write the reconstructed signal for $l = 0, \ldots, N-1$ as

$$x_0(l) = \sum_{k=0}^{L-1} \sum_{r=0}^{p-1} h_0(k) g_0(l + rN) x(-rN - k), \tag{8.1}$$

where $h_0(\cdot)$ and $g_0(\cdot)$ are defined as in Section 2.2.

For a stationary stochastic process $x(l)$ the reconstruction error variance is periodic with period N, and the expression is given by

$$
\begin{aligned}
\sigma_e^2(l) &= E[(x(l) - x_0(l))^2] = \sigma_x^2 - 2E[x(l)x_0(l)] + E[(x_0(l))^2] \\
&= \sigma_x^2 - 2 \sum_{k=0}^{L-1} \sum_{r=0}^{p-1} h_0(k) g_0(l + rN) r_{xx}(l + k + rN) \\
&\quad + \sum_{k=0}^{L-1} \sum_{m=0}^{L-1} \sum_{r=0}^{p-1} \sum_{q=0}^{p-1} h_0(k) h_0(m) g_0(l + rN) g_0(l + qN) \\
&\quad r_{xx}(k - m + (r - q)N), \qquad l = 0, \ldots, N-1. \tag{8.2}
\end{aligned}
$$

For modeling smooth image areas, or, as we only consider separable filter banks, smooth image scan lines, we introduce a slowly varying sinusoid. Thus, we postulate that blocking effects are avoided if the filter bank system is able to reconstruct the sinusoid using lowpass band data only. Following [22], the period of the sinusoid is set to 512 samples, which in our experiments corresponds to the image dimension. This can e.g. be interpreted as an image where the sample values vary smoothly from white (left edge) to black (middle) and to white again (right edge). In this case the input signal is not a stochastic process, and we substitute $r_{xx}(k)$ in Equation 8.2 by the deterministic autocorrelation function of $x(l) = \sin(\frac{2\pi l}{512})$. In Section 8.2.2 we compare coding results of filter banks optimized with and without Equation 8.2 as an error criterion.

In the frequency domain, the issue of blocking effects, when only allowing lowpass data representation, can be seen as an interpolation problem. For a uniform, N-channel critically sampled filter bank and an input sinusoid with normalized frequency f_0, the synthesis lowpass response should suppress the mirror components introduced in the upsampling procedure, as shown in Figure 8.3. The frequencies of the mirror components are given by:

$$f_i^L = \frac{i}{N} - f_0, \qquad i = 1, \ldots, \frac{N}{2}$$

$$f_i^H \;=\; \frac{i}{N} + f_0, \qquad i = 1, \ldots, \frac{N}{2} - 1. \tag{8.3}$$

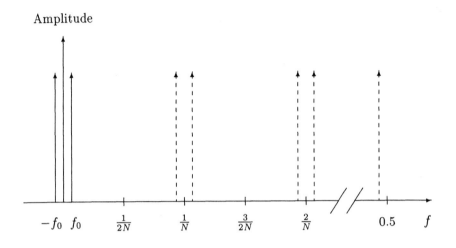

Figure 8.3: *Mirror frequencies of sinusoid with normalized frequency f_0.*

Note that the principle of sinusoid reconstruction depicted in Figure 8.3 is valid for any frequency $f_0 < \frac{1}{2N}$: If the mirror frequencies above $\frac{1}{2N}$ are removed and the original component at f_0 preserved, then the lowpass channel is capable of reproducing that frequency without artifacts. This means that *any* slowly varying signal will be well reconstructed at low bit rates if:

1. The signal does not contain frequencies higher than f_0.

2. The synthesis lowpass response removes all frequencies *between* the mirror components corresponding to f_0, as indicated by dotted arrows in Figure 8.3.

Therefore, the key to a blocking-free reconstruction is to design filter banks where the synthesis lowpass response efficiently removes all frequencies on a region around the critical frequencies $\frac{i}{N}, i = 1, \ldots, \frac{N}{2}$. In Section 8.2, we will relate the frequency responses of the obtained filter banks to their ability to produce a blocking-free reconstruction.

Note that in the 2-channel case, the criterion for removing blocking effects is related to that of Daubechies [37]. In her work the concept of multiple zeros at relative frequency 0.5 is used for regularity purposes. Similar observations were made by Kronander [93], who stressed the necessity of having good suppression of mirror components at relative frequency 0.5, if serious blocking artifacts are to be avoided. In those arguments, only the dc-component was considered. We stress the importance of also providing good attenuation of frequencies close to 0.5. Therefore, some of the available zeros should be placed at frequencies lower than 0.5.

The sinusoid criterion is a generalization in the sense that we are concerned with the reconstruction of smoothly varying signals, rather than constants. In Section 8.2 we design filter banks where the synthesis lowpass responses have good attenuation in a region around the critical frequencies. Kronander's and Daubechies' criterion is only concerned with the filter attenuation *at* the critical frequencies, and the filter banks obtained are not necessarily good at reconstructing slowly varying image areas.

8.1.2 Suitable Quantization Schemes for Higher Bands

As outlined in the beginning of this chapter, we would like to design filter banks specially suited for artifact-free reconstruction at low bit rates. It is therefore important that the smooth image areas are treated as outlined in Section 8.1.1. We must therefore choose a quantization scheme that fulfills the basis restriction assumption in these areas. A suitable scheme is the scalar "dead-zone" quantizer of Gharavi and Tabatabai [84], shown in Figure 4.2. The choice of threshold value t is discussed in Chapter 7, but as a rule of thumb a threshold value equal to the quantization step $(t = \Delta)$ is suitable at low bit rates. This produces a "dead-zone" twice as large as the quantization step.

Vector quantization (VQ) [7] is not easily adapted to our purpose. This method has been successfully applied to subband coding of images [77] by forming vectors across the subbands, but in our setting it is unsuitable due to the lack of individual control over low valued samples in the higher bands. Using VQ, such samples will be coarsely approximated, but not necessarily to zero. Thus, the basis restriction assumption is not fulfilled, and noise artifacts are generated.

Standard bit allocation techniques also present problems, due to the lack of zero representation. As an example, consider the adaptive bit allocation

method explained in Section 6.2.2. The problem with this method is that
the low value samples in higher bands belonging to the 1 or 2 bit classes are
given values different from zero. In fact, only the samples belonging to the
0 bit class are quantized to zero. To obtain zero representation, we must
resort to *level-allocation* rather than bit-allocation. This presents practical
problems since the levels used for quantization usually must be arranged
into corresponding bit representations.

In view of the above mentioned difficulties, we will only use "dead-zone"
quantization when testing filter banks designed for blocking-free reconstruction.

8.2   FIR Filter Banks for Robust Coding

In the previous section we have studied noise artifacts in image subband
coding. In particular, we have given an analysis of the blocking effects
associated with short length unit pulse responses. A criterion for reducing
this type of noise was given.

In the general case, it is not necessary for a (nearly) perfect reconstruction FIR parallel filter bank to have equal filter lengths in all channels. For
our purpose it would be advantageous to allow the *lowpass synthesis response* to extend well beyond that of the other channels, thus giving room
for better suppression of mirror components of low frequency signals, while
assuring that ringing noise from higher bands is controlled.

For obtaining data compression, our filter bank should have good coding
gain. It is desirable to allow for nonunitary filter banks, as substantiated in
Chapter 5.

An important constraint when designing FIR filter banks for low bit
rate image subband coding is that of having linear phase channel responses.
This solves the problem of signal extension at the image borders by allowing
mirror extension, as shown in Chapters 2 and 7. In this method, artificial
high frequency signal generation is avoided by extending the input signal
at the borders by mirror reflection. This is not the case for the circular
extension method, where the signal is extended using the opposite end of the
data, thus possibly creating an artificial signal step. Furthermore, from the
gain formula, Equation 5.36, we see that the filter phase does not *directly*
enter the gain expression. This is reflected in optimization experiments:
We have never obtained nonlinear phase filter responses with better gain

than corresponding linear phase filters of equal length. Note also that the filtering complexity is reduced by 50% (at the cost of some storage memory) when using linear phase filters, since the associated unit pulse responses are symmetric or antisymmetric.

When choosing frequency partitioning and filter lengths for an SBC system, we must consider both stationary as well as nonstationary input signals. For a stationary signal having nonwhite power spectrum, the coding gain increases with a partition refinement [10], see Figure 5.4. However, fine frequency splitting invariably leads to long unit pulse responses (see Appendix B), which are unsuitable for the representation of real-world data [95] because they result in ringing artifacts in the presence of noise. From a gain point of view, the loss by using uniform, 8 band partitioning is small for the AR(1), $\rho = 0.95$ process as was observed in Figure 5.4. In this section we will limit the discussion to the study of uniform filter banks with $N = 8$ channels. This is mainly motivated by the success of uniform filter banks like the DCT and the LOT. As we saw in Chapter 7, these filter banks exhibit a limited amount of ringing noise, but suffer from blocking effects. We shall try to improve the coding performance in that respect.

To summarize, we need a way to design an FIR, parallel, linear phase, uniform filter bank having (almost) perfect reconstruction, good suppression of blocking effects as well as good coding gain. In addition, the synthesis responses should be kept short in order to avoid serious ringing noise. In Section 2.5 we presented a practical approach due to Nayebi et al. [20, 21]. In the rest of this section we shall see how this method for designing nonunitary filter banks can be used for performance optimization.

8.2.1 The Error Function

Having presented the framework of a general method for designing (almost) perfect reconstruction filter bank systems (Section 2.5), we now want to incorporate suitable error terms for minimization: High coding gain and removal of blocking effects while keeping the *synthesis* unit pulse responses short in order to avoid extensive ringing noise. The latter objective is the reason for putting the synthesis filter coefficients into the $\mathbf{A}$ matrix in Equation 2.47. Remember that the $\mathbf{H}$ matrix is dependent on the $\mathbf{A}$ matrix according to Equation 2.54 and as such it does not contain free parameters in the optimization algorithm. Thus, if we want the synthesis channels to have filter lengths less than $L = pN$, this must be specified in the $\mathbf{A}$ matrix.

In general we let all unit pulse responses for the analysis channels have full length L.

A suitable error function for our purposes is

$$\varepsilon = v_P \varepsilon_P + v_G \varepsilon_G + v_B \varepsilon_B, \tag{8.4}$$

where the error terms are included to account for (ε_P) *perfect reconstruction*, (ε_G) *coding gain*, and (ε_B) *blocking effects*. The respective weight factors v_P, v_G, and v_B will be discussed in the following section.

The coding gain is defined as

$$\varepsilon_G \overset{\triangle}{=} \frac{1}{G_{SBC}} = \frac{\left[\prod_{n=0}^{N-1} \sigma_{y_n}^2\right]^{\frac{1}{N}}}{\sigma_x^2}, \tag{8.5}$$

where G_{SBC} is the subband coding gain measure presented in Chapter 5. To simplify the filter bank optimization, the gain contribution of the quantizer performance factors, $\frac{\epsilon_x}{\epsilon_{yg}}$, is assumed independent of the filter responses. This is evident from Equation 8.5, where $\frac{\epsilon_x}{\epsilon_{yg}}$ is set to unity. When quoting coding gains, we shall be referring to the gain due to the variances only. This is an exact result for a Gaussian input source.

The blocking effects term is defined as

$$\varepsilon_B \overset{\triangle}{=} \frac{1}{N} \sum_{l=0}^{N-1} \sigma_e^2(l), \tag{8.6}$$

where $\sigma_e^2(\cdot)$ is defined in Equation 8.2.

As shown in Chapter 5, if all synthesis filters have unit norm the standard bit-allocation formula can be used also for nonunitary filter banks. All filter banks presented in this chapter have this property.

Finding explicit expressions of the error functions and their derivatives, i.e. as functions of the analysis and synthesis filter taps, is a laborious task and omitted here. Full details can be found in [22].

Linear Phase

It is well known [14, 15] that in coping with image boundaries, the mirror extension method offers superior coding results to that of the circular extension at low bit rates. In Section 2.4 we saw that when using mirror extension, it is necessary that the channel unit pulse responses are all linear phase. In our design scheme it is therefore important to ensure that this constraint is satisfied.

In [96] it is shown that the eigenvectors of an autocorrelation matrix always have linear phase. Furthermore, exactly half of the vectors have even symmetry and the other half have odd symmetry. This is also the case for the basis functions of the DCT transform, where the first vector, corresponding to the lowpass filter, has even symmetry. The symmetry of the other vectors then alters between odd, even, odd, even, etc. In our design scheme we force this type of linear phase symmetry onto the filter banks. For $N = 8$ the filter length must be an even number since $L = pN$.

A nice feature of Equation 2.54 is that the analysis filters are guaranteed to be linear phase if the synthesis filters are linear phase, regardless of whether we have perfect reconstruction or not [22]. Choosing linear phase responses also means reducing the number of free coefficients in the gradient search by 50%.

8.2.2 Optimization and Parameter Tuning

In previous sections we have described a very general procedure for designing uniform, perfect reconstruction, FIR, parallel filter banks. The choice of design parameters will influence the performance of the resulting filter bank.

This section gives an overview over the multitude of design considerations, and presents various filter banks optimized for low bit rate coding. As before, a uniform 8×8 band splitting scheme is used.

Initial Values

The gradient search algorithm needs initial values of the synthesis filter coefficients constituting the free parameters $\{g_n(l)\}$. For faster convergence and for reducing the risk of falling into a local minimum, the initial values should correspond to a proper filter bank system, i.e. all filters should have reasonable bandpass characteristics and the total system should be close to perfect reconstruction. In most experiments we have used the $N = 8$ channel DCT as starting point for the iterations. When $L > 8$ (as is the case in all our experiments) the initial filter coefficients outside the 8-tap center are set to zero.

Weight Factors

The total error function consists of 3 terms: (ε_P) *perfect reconstruction*, (ε_G) *coding gain*, and (ε_B) *blocking effects*. As shown in Equation 8.4, each error term is multiplied by a weight factor.

The first weight factor should be chosen such that the resulting filter bank can reconstruct images at an SNR of approximately 50 dB in the absence of quantization noise. For our purpose, 50 dB will in practice correspond to error free *visual* reconstruction, and a higher weighting factor will simply decrease the freedom we have to minimize the other terms. This weight factor is found by a systematic trial and error procedure after the other factors have been fixed. During optimization we want to obtain a reasonable trade-off between coding gain and blocking effects. The latter objective is considered crucial for subjective coding quality, and in practice the weight factors for these terms should be chosen to allow for blocking-free reconstruction.

Filter Lengths and Naming Conventions

As explained in Section 8.2.1, we want to restrict the length of the synthesis unit pulse responses, while allowing the analysis responses to have full length L. In practice this means using as short synthesis filter lengths as possible, while avoiding blocking effects and serious loss in coding gain. The former objective is more easily achieved if we allow the synthesis *lowpass* response to have full length L, as explained in the introduction to Section 8.2.

We also stated that high frequency information is better represented using short unit pulse responses whereas low frequency information should be represented using long responses. This suggests that the filter length should decrease with the channel number. Figure 8.4 shows a synthesis matrix containing a plausible choice of responses for the synthesis filter bank. The nonzero filter coefficients are shown as black lines.

In this example the responses of the 5 highest frequency bands in the synthesis filter bank have lengths equal to the upsampling factor $N = 8$, thus corresponding to a pure square transform decomposition for those bands. This is easily observed by noting that these responses do not overlap. Also note that all responses, within each block of 8, are centered around the same point of symmetry. If we apply a filter bank with lengths similar to that

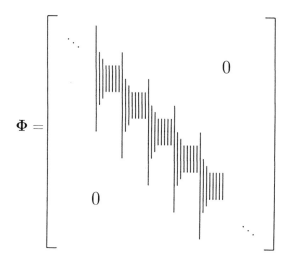

Figure 8.4: *Synthesis responses with varying length.*

of Figure 8.4, and use DPCM with high bit rate to quantize the LP-LP band, the long lowpass response will not give rise to extensive ringing noise. Anticipating the optimization results in Section 8.2.3, we shall see that the use of the blocking effects criterion leads to very smoothly decaying synthesis lowpass unit pulse responses with no ripple. This is very beneficial from a ringing noise point of view because it means that a step signal like that of Figure 8.1 will not contain ringing amplitudes after being filtered through such a synthesis lowpass filter.

We examine the coding merits of several choices of synthesis filter lengths. Table 8.1 shows the 4 main configurations used. In all cases the analysis filters have full length L.

It is seen that configuration I allows for full filter lengths L, whereas configurations II to IV are increasingly restrictive – the latter is indeed close to a pure transform. The coefficients of the higher bands are centered around the point of symmetry defined by the lowpass channel.

Suitable values for L are 32, 24 and 16. Recall that L is a multiple of the upsampling factor N. Higher values (40 and up) are not considered because of the associated ringing noise.

We adhere to the following naming convention: A main configuration X, with synthesis lowpass channel of length L is referred to as configuration LX. For reference, the synthesis responses depicted in Figure 8.4 correspond

	Configuration			
Channel	I	II	III	IV
0	L	L	L	L
1	L	16	12	8
2	L	12	10	8
3	L	8	8	8
4	L	8	8	8
5	L	8	8	8
6	L	8	8	8
7	L	8	8	8

Table 8.1: *Main configurations for synthesis filter lengths.*

to configuration 32II.

Optimization

The error function of Equation 8.4 is minimized by application of the *Broyden-Fletcher-Goldfarb-Shanno* (BFGS) gradient search method implemented in [97].

The disadvantage of using gradient search algorithms is that we have to "search in the dark" for good solutions. In this investigation we can not claim definitive optimality.

We try, however, to make as extensive searches as possible, and this is an outline of the strategy:

1. Test the necessity of the error term for blocking effects.

2. Choose a blocking-free filter bank, and check whether the value of the perfect reconstruction error weight v_P influences the result or not. A lower value might give lower error terms for coding gain and/or blocking effects at the cost of reduced reconstruction quality in the absence of quantization noise.

3. For the same filter bank configuration, adjust the values of v_G, v_B and test whether the trade-off between coding gain and blocking effects can be improved, i.e. try to increase the coding gain while keeping the blocking effects term below a critical level.

4. Test whether the design procedure is sensitive to the choice of initial values.

Blocking Effects

To verify the usefulness of the basis restriction model introduced in Section 8.1.1, we optimize the coefficients with and without the blocking effects error term in the error function. Experience shows that, in general, the error term for blocking effects, ε_B, should be included for filter bank optimization for $L = 24$ or less. This becomes increasingly more important when using the length-restrictive configurations II, III and IV. Table 8.2 shows the obtained coding gains with and without the blocking effect term for some filter bank configurations.

Configuration	32I	32IV	24IV	16I
Without blocking effects term	9.75	9.38	9.25	9.63
With blocking effects term	9.75	9.37	9.25	9.61

Table 8.2: *Obtained coding gain (dB) with and without the blocking effect term. An AR(1), $\rho = 0.95$ signal model is assumed.*

We observe that very little coding gain has been traded for the sake of reduced blocking effects. Indeed, for all configurations examined the reduction is limited to 0.03 dB.

The coding performance of the designed filter banks is compared with that of Malvar's 16-tap LOT [52] and the DCT. Several 512×512 pels grey tone images are coded at bit rates around 0.4 bits per pel. The method of quantization is the same as was used in Chapter 7, with $t = \Delta$. To illustrate the resulting blocking effects, we show mesh plots of a 32×32 pels smooth image section: "Lenna's shoulder", a subsection of the image "Lenna". The full original image is shown in Appendix D, Figure D, where the subsection is indicated with a white frame. The same image section is shown as a mesh plot in Figure 8.5. The results are shown in Figure 8.6 and 8.7 for filter banks designed without and with the blocking effects term, respectively.

Note the cancellation of the "granular noise" in the original image section. This is a consequence of the applied threshold in the quantization procedure. The loss of high frequency components in this region is not serious in terms of visual quality. A complete rendition of these high frequency

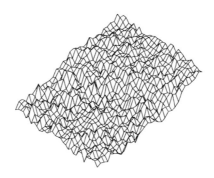

Figure 8.5: *Original* 32×32 *pel shoulder region from "Lenna" image.*

components would require excessive bit rate. However, this type of granular noise could be added after reconstruction if providing the statistical noise parameters as side information, thus improving the natural appearance of the image.

From Figure 8.6 we see that all filter banks, with the exception of configuration 32I, exhibit blocking effects of varying degrees. For 32IV it is barely noticeable, but it increases significantly for the other filter banks, although with different characteristics: The 16I filter bank reconstruction is somewhat rugged with "valleys and hills", whereas the 24IV reconstruction exhibits clear blockiness. The DCT exhibit "true" blockiness in the sense that the reconstructed signal has large discontinuities, whereas the LOT offers a "valleys and hills" landscape.

Figure 8.7 demonstrates the merits of the suggested compound optimization criterion. None of the reconstructed images exhibit blocking artifacts. With reference to Table 8.2, we can conclude that the coding gain traded for reduced blocking effects proved to be an excellent investment.

We have also conducted experiments with larger values of f_0, the normalized frequency of the sinusoid model. The 32II configuration was re-optimized with $f_0 = \frac{1}{256}$. The resulting filter bank coefficients were practically unchanged and therefore no difference in blocking effects resulted. We therefore conclude that $f_0 = \frac{1}{512}$ is sufficient for avoiding blocking artifacts in uniform, $N = 8$ channel filter banks.

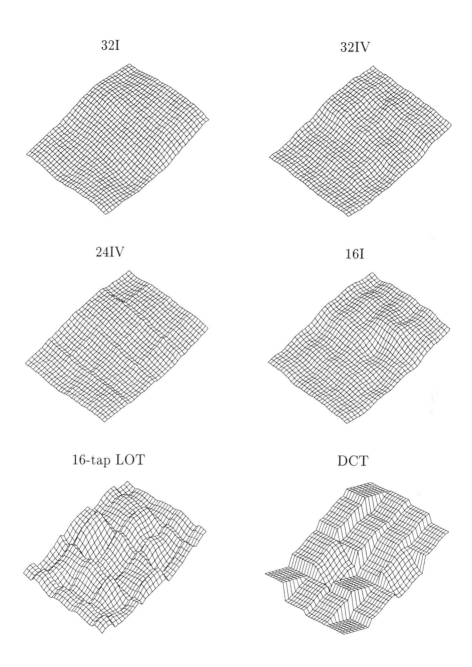

Figure 8.6: *Reconstructions* without *blocking effects optimization.*

32I   32IV

24IV   16I

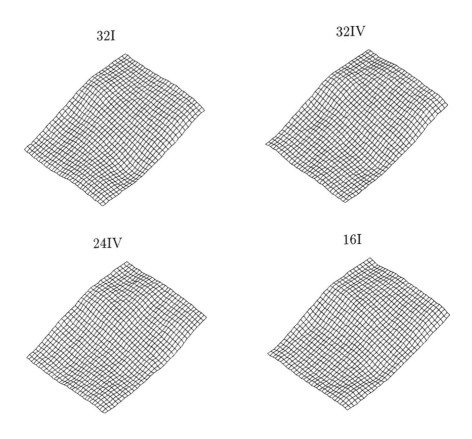

Figure 8.7: *Reconstructions* with *blocking effects optimization.*

Coding Gain vs. Blocking Effects

Comparing the obtained coding gains with the results of Cassereau et al. [76] and Malvar [52, 98, 75], the highest coding gains claimed for $N = 8$ channels were 9.24 dB and 9.51 dB for 16 and 32 tap LOT/ELTs, respectively. Two of the obtained filter banks, the 32I (9.75 dB) and 24I (9.64 dB) have even higher gain than an 8 channel ideal brick wall uniform filter bank: 9.62 dB (Figure 5.4).

We attribute the improved coding gains to the increased amount of freedom available through use of nonunitary filter banks. As discussed in Chapter 5, the concept of half-whitening explains the higher coding gains of nonunitary filter bank systems.

In [22], a systematic search was conducted for improving the trade-off between blocking free reconstruction, maximum coding gain, and sufficiently precise reconstruction in the absence of quantization noise. This was done using the 32II configuration. The resulting filter bank, denoted 32II_opt, has a coding gain of 9.63 dB, typical reconstruction error in the 50-55 dB range, and the blocking effects error is in a critical range. It was therefore tested for visual evaluation. Figure 8.8 shows the obtained image subsection.

32II_opt

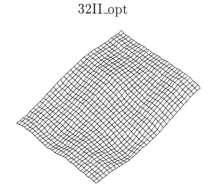

Figure 8.8: *Reconstruction detail using 32II_opt filter bank.*

Compared with the reconstructions shown in Figure 8.7, this reconstruction suffers from a very slight ruggedness: It is in a critical area of smoothness. However, looking at the decoded image on the computer screen, it is impossible to detect any systematic noise artifacts in the smooth image areas, and we conclude that the blocking effects errors are visually acceptable.

We repeat at this point that all optimization experiments have been conducted with the DCT transform as the starting point for the iterations. In [20, 21] it is claimed that the design method is insensitive to the choice of initial values, and this was also confirmed in [22]. It should also be emphasized that in these optimization experiments there is a risk of falling into a local minima.

8.2.3 Time and Frequency Responses

We now focus on the time and frequency characteristics of the optimized filter banks. In this investigation we are particularly interested in relating the improved coding gain and absence of blocking effects to the filter characteristics. It is therefore natural to compare to the 16-tap LOT as well as the 8 tap DCT transform, and point out similarities and differences. For a fair comparison with the LOT we choose the 16I configuration, which has the same filter response lengths. We also investigate the fine-tuned filter bank: 32II_opt.

Time Domain Characteristics

The new filter banks are nonunitary. It is therefore interesting to highlight the differences between corresponding analysis and synthesis responses. In Figure 8.9 the unit pulse responses of the four lowest frequency band channels of the 32II_opt filter bank are shown, using equal scaling.

There is a dramatic difference between the analysis and synthesis lowpass responses: The analysis response has two abrupt transitions, and while it tapers off to zero at both ends, the decay is not as smooth as that of the synthesis response. The smoothness of the synthesis response is related to the blocking free reconstruction properties of this filter bank, and is in accordance with Malvar's approach [75], in which he tried to design the basis vectors of the LOT to decay smoothly to zero. An additional advantage of the smooth, monotonously decaying synthesis lowpass response, is that it does not produce ringing noise. A signal edge reconstructed with this response, will be "smeared out", but without any over- or undershoots as in Figure 8.1. This is important, since we have allowed for a relatively long lowpass response. The synthesis responses, excluding the lowpass response, have limited lengths, as given by Table 8.1. This is, to a lesser extent, also the case for the analysis responses, and becomes more pronounced for the higher frequency band responses. In addition we see that corresponding analysis and synthesis responses also become more similar in terms of shape.

Similar plots for the 16I filter bank and the LOT are shown in Figure 8.10. The same comments as those pertaining to the 32II_opt filter bank can be made here, with the exception that all responses have full length. The advantage of allowing for nonunitary filter banks becomes evident when looking at the lowpass responses. The analysis response has very large end-tap values, and would be a disaster in terms of blocking effects if used in the synthesis stage. Compared to the LOT, it is seen that the end taps of the lowpass LOT response have significantly larger amplitude than the end taps of the corresponding 16I synthesis response.

32II_opt analysis 32II_opt synthesis

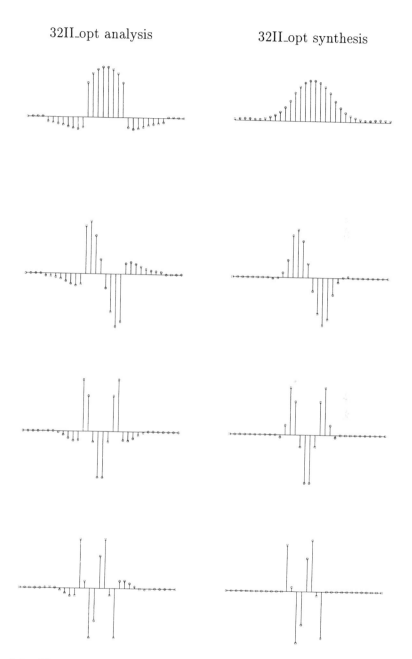

Figure 8.9: *Unit pulse responses of the four lowest frequency band channels for 32II_opt; lowpass at top. The synthesis responses are time-reversed to provide easy comparison with the corresponding analysis responses.*

16I analysis　　　　　　16I synthesis　　　　　　16-tap LOT

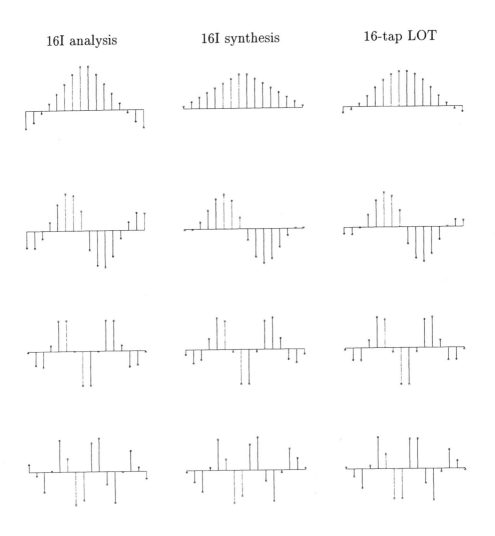

Figure 8.10: *Unit pulse responses of the four lowest frequency band channels for filter banks 16I and 16-tap LOT; lowpass at top. The synthesis responses are time-reversed to provide easy comparison with the corresponding analysis responses.*

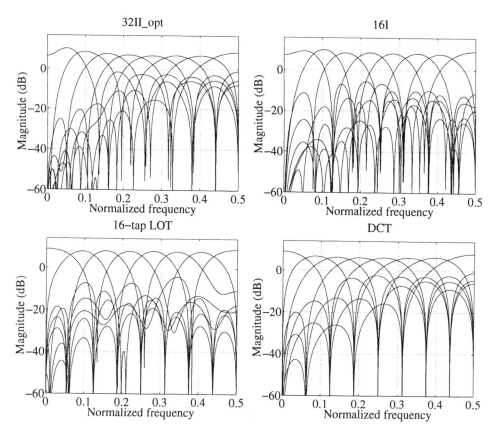

Figure 8.11: *Analysis frequency responses.*

Frequency Domain Characteristics

The absence of dc leakage of a subband coding system is generally recognized as a crucial feature, regardless of whether AR(1) coding gain, simultaneous space-frequency localization, or other criteria are used as performance measures for the filter bank. The frequency responses of the higher frequency band analysis channels should provide near perfect removal of the dc component – failure to do so will lead to high signal energy in the corresponding bands, thus degrading the coding performance. In Figure 8.11, we compare the 32II_opt and 16I analysis responses with those of the LOT and the DCT.

None of the filter banks exhibit any dc leakage, but the LOT and DCT

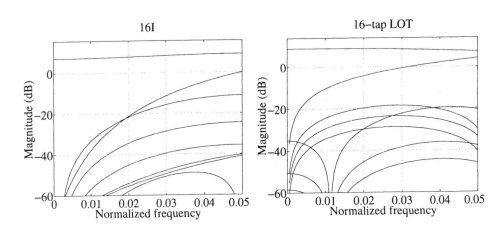

Figure 8.12: *Analysis frequency responses at low frequencies.*

have generally more leakage of low frequencies into the higher frequency bands. This is particularly evident if we compare the first bandpass response for the 16I and the LOT. The 16I response has a narrower shape and will remove much more low frequency signal power due to increased suppression of frequencies below 0.05.

Detailed plots of this region are shown in Figure 8.12. The 16-tap LOT shown here has a coding gain of 9.24 dB, whereas the 16I filter bank has a coding gain of 9.61 dB – slightly below that of the 16-tap filter bank presented in Chapter 5. This is due to the blocking effects criterion being used when designing the 16I filter bank.

The synthesis responses of 32II_opt and 16I, Figure 8.13, differ considerably from those on the analysis side. The lift in the lowpass response is compensated for by a corresponding decay, and the higher band responses exhibit serious dc leakage (i.e. dc leakage if we were to use it on the *analysis* side of a subband coder). This is irrelevant since the signal decorrelation is performed at the analysis side, and it allows for more freedom for other error terms, since no zeros in the amplitude transfer function have to be used for this purpose.

The blocking effects phenomenon is strongly related to the lowpass responses. We therefore plot these separately in Figure 8.14, using larger scale for the magnitude axis.

The synthesis responses of 32II_opt and 16I clearly show how blocking effects are avoided by removal of the mirror frequencies, as explained in

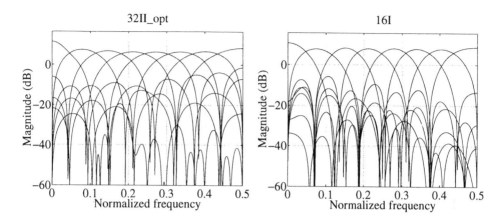

Figure 8.13: *Synthesis frequency responses.*

Section 8.1.1. The mirror frequencies are located in the regions limited by

$$f_i^L = \frac{i}{8} - f_0, \qquad i = 1, 2, 3, 4,$$

$$f_i^H = \frac{i}{8} + f_0, \qquad i = 1, 2, 3, \qquad (8.7)$$

where f_i^L and f_i^H denote the lowest and highest normalized frequencies for location number i. For blocking free reconstruction of sinusoids with frequency less than f_0, the lowpass response should suppress all frequencies between f_i^L and f_i^H for the given values of i.

In the filter bank optimization, we used a sinusoid of period 512. This corresponds to a normalized frequency of $f_0 = \frac{1}{512} \approx 0.002$, thus we want suppression of each interval of length 0.004, centered about the frequencies $0.125, 0.25, 0.375$ and 0.5.

The 32II_opt and 16I filter responses clearly have better suppression at these frequencies, compared to the LOT and the DCT. Note that in most cases, multiple, closely spaced transmission zeros are used to obtain a wide stopband for 16I and 32II_opt. Still, the LOT and the DCT do have zeros at the critical locations and this is a direct consequence of the absence of dc leakage for these filter banks.

A nice feature of the blocking effects removal criterion (Equation 8.2) is that it tends to generate smoothly decaying lowpass synthesis responses with no ripple, as we saw in Figures 8.9 and 8.10. This leads to smoothing,

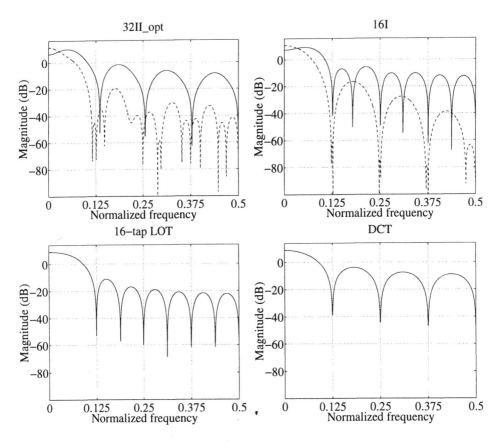

Figure 8.14: *Lowpass frequency responses: Solid=analysis, dotted=synthesis. The grid indicates the critical mirror frequencies on the horizontal axis.*

rather than ringing noise when coding images at low bit rates. To see the connection, consider a rectangular pulse $w(n)$ with height 1 and length K. The corresponding discrete-time-fourier-transform is

$$
\begin{aligned}
W(e^{j2\pi f}) &= \sum_{l=0}^{K-1} e^{-j2\pi fl} = \frac{1 - e^{-jK2\pi f}}{1 - e^{-j2\pi f}} \\
&= \frac{\sin(K\pi f)}{\sin(\pi f)} e^{-j(K-1)\pi f}.
\end{aligned} \tag{8.8}
$$

Setting $K = 8$, this response has zeros at frequencies

$$f_k = \frac{k}{8}, \qquad k = 1, \ldots, 4, \qquad (8.9)$$

which are the critical center frequencies in the blocking effects removal criterion. It follows that a response having multiple, say p, zeros at each of these frequencies, can be obtained by convolution of p rectangular pulses. For $p = 2$ a triangle of length 15 results, which is very close to the response of the lowpass synthesis response of filter bank 16I, Figure 8.10, top middle. For $p = 4$ it resembles a bell shape of length 29, not dissimilar to the lowpass synthesis response of filter bank 32II_opt, Figure 8.9, top right.

For $K = 8$, $|W(e^{j2\pi f})|$ is identical to the lowpass response of the 8-channel DCT transform, Figure 8.14, bottom right. Still, the DCT does exhibit blocking effects, and this simply results from the insufficient attenuation provided by single zeros at the critical frequencies, or, equivalently, from the abrupt transition to zero at each end of the lowpass unit pulse response.

8.2.4 Coding Examples

In previous subsections of this chapter we have made an effort to show how to design filter banks suitable for low bit rate image coding. To justify our efforts, we conclude this section with coding experiments similar to those of Chapter 7. Note that lowpass band DPCM quantization is used in all experiments for a fair comparison with the unitary filter banks (DCT and LOT). This is not in line with the assumption of PCM quantization of *all* bands made when optimizing the filter banks of this chapter. It is to be expected that the use of DPCM in the lowpass band calls for special attention when designing the lowpass filters of the filter bank. An investigation of this topic follows in Section 8.2.5.

Ringing Noise Evaluation

For evaluating the ringing noise resulting from a filter bank it is natural to compare with the DCT since it exhibits a minimum of this noise type[1] . It is to be expected that some of the filter banks, namely those belonging to the II, III and IV configurations of Table 8.1, have similarly attractive features.

---

[1] Assuming an $N = 8$ channel, uniform filter bank.

As pointed out in Section 8.1, ringing noise is generated in regions of sharp transitions. Furthermore, it is masked by textures and detailed regions. For evaluation we use the image subsection first shown in Figure 7.2.

Figure 8.15 depicts the coding results obtained when using "dead-zone" quantizer with threshold level equal to the quantization step size ($t = \Delta$), in conjunction with several filter banks. The bit rates are in the range 0.43 to 0.46 bits per pel.

The ringing noise is particularly visible on both sides of the vertical top mast, particularly for the 32I filter bank, but also for 16I. The other filter banks have a ringing noise quite similar to that of the DCT (which instead is haunted by blocking effects). This is a direct consequence of the short length of the higher band synthesis channels of these filter banks. In these coding experiments, all filter banks belonging to the II, III and IV configurations (except 16IV) have generated a similar amount of ringing noise.

Coding Gain of Real-World Images

When designing filter banks for low bit rates we applied an error term for theoretical coding gain. For filter bank 16I, a coding gain of 9.61 dB was obtained, whereas the 16-tap LOT has 9.24 dB. The gain was computed assuming the input signal to be an AR(1) process with correlation coefficient of $\rho = 0.95$. The success of this procedure therefore rests on the validity of the signal model, as well as quantization noise assumptions. As we use DPCM for the LP-LP band, in conjunction with low bit rate "dead-zone" quantization of the higher bands, neither assumption is fulfilled.

In Figure 8.16 we have plotted obtained SNR and bit rate for these two filter banks, with varying quantization step size and $t = \Delta$.

For the "Lenna" image, the curve for the 16I filter bank is approximately 0.2 dB above that of the LOT, whereas for the "Kiel" image the curves are practically identical at low bit rates, with a slight preference for the LOT at high bit rates. For "Boat" the 16I filter bank outperforms the LOT with approximately 0.1 dB.

These experiments demonstrate the pitfalls of image signal modeling. The "Lenna" image has lowpass character (i.e. smooth regions and few details) and is therefore reasonably well modeled with an AR(1), $\rho = 0.95$ spectrum, whereas the "Kiel" image is extremely detailed with significant high frequency content in the bottom half of the image. "Boat" is somewhere in between these two in terms of its frequency spectrum. These coding results

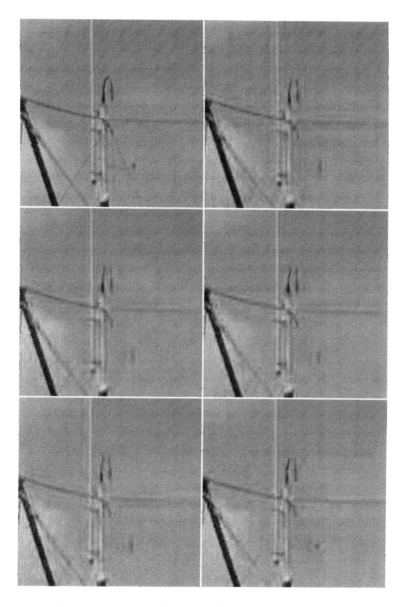

Figure 8.15: *Ringing noise: Original (top left), 32I (top right), 32II_opt (middle left), 32IV (middle right), 16I (bottom left), and DCT (bottom right).*

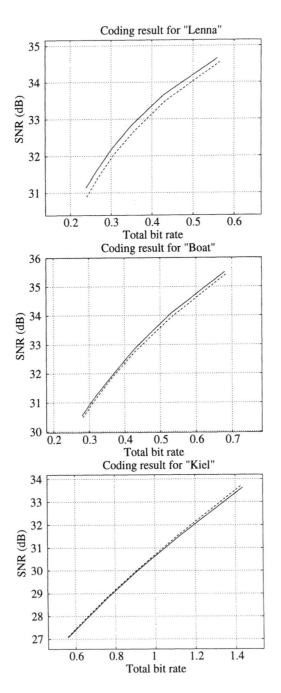

Figure 8.16: *SNR vs. bit rate: 16I in solid line, 16-tap LOT in dotted line.*

are in accordance with our investigation of the frequency responses of the two filter banks in Section 8.2.3, where it was pointed out that the higher band responses of the 16I filter bank had better suppression of lowpass energy. It will therefore perform better for images that are of lowpass character.

Figure 8.17: *Coding result using filter bank 16I.*

The coding gain measure is justified only if used in conjunction with other design criteria: The *visual* performance of the 16I filter bank is better than that of the LOT due to the absence of blocking effects. This is demonstrated in Figure 8.17 at a total bit rate of 0.4 bits per pel with $t = \Delta$. Compare to Figures 7.6 and 7.7 in Chapter 7. The 16I filter bank has blocking-free reconstruction similar to the FIR and IIR filter banks, but with considerably less ringing noise.

8.2.5 Subband Coding and DPCM

In Chapter 5 we concluded that the optimal AR(1) coding gain of nonunitary filter banks is achieved by allowing half-whitening filters in each filter bank channel. In the derivation of the subband coding gain formula we assumed

PCM quantization in each subband. The common practice, however, is to quantize the lowpass band using *closed-loop* DPCM in order to exploit intraband correlation. This is done in all coding experiments in this book.

The $N = 8$ channel filter banks obtained in this chapter have a small lift in the analysis lowpass responses. When used in conjunction with closed-loop DPCM quantization of the LP-LP band, suboptimal performance results because some signal correlation is removed from the LP-LP signal prior to DPCM coding.

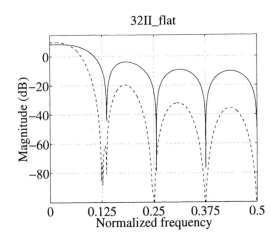

Figure 8.18: *Lowpass responses: Solid=analysis, dotted=synthesis.*

An investigation on the significance of the nonflat frequency analysis lowpass response is in order. A filter bank is redesigned with restrictions incorporated into the gradient search error function. To obtain a flat response we force the derivative of the lowpass analysis response to be close to zero in a small frequency region. The 16I configuration does not contain enough free parameters for this purpose. This is evident from Figure 8.14, which depicts the 16-tap LOT and the 16I lowpass frequency responses. The 16I synthesis response avoids blocking effects by spending 2 transmission zeros in the vicinity of each mirror frequency component, as given by Equation 8.9. The transmission zero at relative frequency 0.5 is real whereas the others are complex, making a total of 15 zeros in the complex plane. If the analysis lowpass response is made flat in the passband, the synthesis response must also be flat, and this calls for additional zeros in the passband region, thus making it similar to the 16-tap LOT lowpass response. We therefore select the 32II configuration for optimization.

The lowpass filter response was made flat by simply adding an additional error term to the error function. It was sufficient to force the derivative of the lowpass frequency response, $H_0'(f)$, to zero for a few discrete frequencies [22]. The obtained lowpass frequency responses are shown in Figure 8.18. Compared with the 32II_opt responses in Figure 8.14, the new filter bank, denoted 32II_flat, has better attenuation of mirror frequencies. This is just a consequence of our fine-tuning of the 32II_opt filter bank. The coding gain is reduced from 9.63 dB to 9.50 dB.

32II_flat analysis   32II_flat synthesis

Figure 8.19: *Lowpass unit pulse responses.*

The lowpass unit pulse responses are shown in Figure 8.19. When compared to the 32II_opt responses in Figure 8.9, we notice how the restriction for flat lowpass response has imposed a sinc-like shape on the synthesis unit pulse response. This follows from Fourier relations. For our purpose the slight undershoot in the synthesis response is of some concern because it might give rise to more ringing noise.

In Figure 8.20 we have plotted obtained SNR and bit rate for the 32II_opt and the 32II_flat filter banks. The same quantization procedure as before was used, with 5 bits per pel when quantizing the lowpass band with DPCM.

For "Lenna" and "Boat" a slight improvement in SNR results, whereas for the "Kiel" image, the curves are practically identical. Visual inspection of the decoded images confirms that no difference can be detected, including the ringing noise level.

The filter coefficients of some of the filter banks designed in this chapter are given in Appendix C.

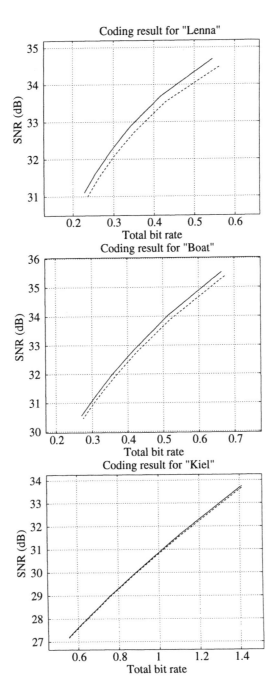

Figure 8.20: *SNR vs. bit rate: 32II_flat in solid line, 32II_opt in dotted line.*

8.3 Subband Coding with Adaptive Filter Banks

In the previous section we considered the use of relatively short unit pulse responses in a filter bank coding system. The use of as short filters as possible while still avoiding blocking effects was emphasized. It was tacitly understood that the filter responses were fixed, i.e. the whole image was decomposed into subband signals using the same responses or basis vectors for different image parts. A different approach is that of varying the filter response lengths within the image, thus giving the notion of *spatially adaptive* filter banks.

Nayebi et al. [99] appear to be the first to describe a locally adaptive analysis/synthesis filter bank system. The treatment in [99] is done in the context of a two band system employing FIR filters of fixed length. A different approach was developed in [100] using adaptive space-frequency tilings. This is a technique for allowing space-varying frequency splitting. An example of this was shown in Figures 3.36 and 3.37.

In the following we show that the IIR filter banks described in Chapter 3 can easily be made spatially adaptive without compromising their perfect reconstruction property. This rather surprising fact is used to simultaneously minimize ringing artifacts and blocking effects in a subband coder. This is achieved by varying the effective impulse response length dynamically, depending on the nature of the input signal segment to be filtered. Here the adaption is done in a blockwise manner implying that for each image block, the filter bank whose characteristics are best suited to the image content of the block is selected. To reconstruct the compressed image, the decoder must therefore receive side information indicating which filter bank was used in the analysis stage for each block. We show that the increase in bit rate is negligible.

8.3.1 Making IIR Filter Banks Adaptive

In Figures 3.15 and 3.16 we presented a general two-channel filter bank structure for PR QMF filter banks. A two-band perfect reconstruction analysis/synthesis system based on IIR all pass filters is obtained by substituting

$$P_i(z) = A_i(z) = \frac{a_i + z^{-1}}{1 + a_i z^{-1}}, \qquad i = 0, 1, \tag{8.10}$$

and in the synthesis filter bank the polyphase filters are given by

$$\frac{1}{P_i(z)} = \frac{1}{A_i(z)} = A_i(z^{-1}), \qquad i = 0, 1. \tag{8.11}$$

Remember that the filters in the synthesis part are unstable and anti-causal. As discussed in Section 3.4.3 and Appendix A we can implement this filtering operation by using the causal filters $A_i(z)$ on reversed versions of the subband signals.

In making the filter banks adaptive, we allow the coefficients a_i to vary. This is indicated by referring to the filter coefficients as $a_i(l)$. The situation to be analyzed when the cancellation of the crossover terms of Figures 3.15 and 3.16 is taken into account, is depicted in Figure 8.21.

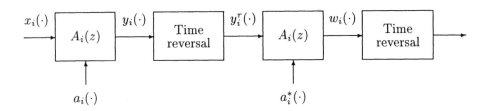

Figure 8.21: *Adaptive all pass filtering.*

The signal domain operations on a sequence $x_i(l)$ by a first order all pass filter with varying coefficients is expressed as

$$y_i(l) + a_i(l)y_i(l-1) = a_i(l)x_i(l) + x_i(l-1). \tag{8.12}$$

In the synthesis stage, the reversed sequence $y_i^r(m) = y_i(-m)$, is filtered through the same filter, but with modified coefficients $a_i^*(m)$:

$$\begin{aligned} w_i(m) + a_i^*(m)w_i(m-1) &= y_i^r(m-1) + a_i^*(m)y_i^r(m) \\ &= y_i(-m+1) + a_i^*(m)y_i(-m). \end{aligned} \tag{8.13}$$

Setting $a_i^*(m) = a_i(-m+1)$ and $l = -m+1$ we obtain

$$w_i(-l+1) + a_i(l)w_i(-l) = y_i(l) + a_i(l)y_i(l-1). \tag{8.14}$$

Combining Equation 8.12 and 8.14 we get an expression for the relation between the input and output signals, given by

$$w_i(-l+1) + a_i(l)w_i(-l) = a_i(l)x_i(l) + x_i(l-1). \qquad (8.15)$$

Now if there exists an integer p such that

$$w_i(-p) = x_i(p), \qquad (8.16)$$

it follows that

$$w_i(-l) = x_i(l), \qquad l < p. \qquad (8.17)$$

In practice the input signal $x_i(l)$ is finite: $l \in \{0, \ldots, K-1\}$, and for implementing circular convolution it suffices to extend the index range to $\{-K, 2K-1\}$. In fact, for the filter banks in this chapter it is sufficient to extend by 10-15 indices to allow the filter transients to fall to zero. After filtering, both $x_i(\cdot)$ and $w_i(\cdot)$ will fall to zero outside this region, thus fulfilling the previously mentioned assumption (Equation 8.16).

In essence we have shown that for an arbitrary variation of the filter coefficients in our two-band, IIR filter bank, perfect reconstruction is maintained [101].

In the following we limit the discussion to first order allpass filters, mainly because of their simplicity, but also because is is known that increased order does not result in significantly better coding results in the nonadaptive case [51].

8.3.2   Application to Image Subband Coding

As pointed out several times in this book, the problem of transform coding is blockiness, whereas in traditional subband coding, the main problem manifests itself as ringing noise in the vicinity of sharp edges. In the following we show how these artifacts can be simultaneously reduced by the use of the adaptive scheme described in the previous section.

In [51] an extensive study of IIR filter banks for image coding was presented. A tree-structured, 8×8 filter bank, giving a total of 64 spatial frequency bands was used. Using fixed filter banks for the whole image it was found that, in terms of SNR, the IIR image subband coders had similar performances as long as the stop band attenuations of the channels were more than 20 dB. In terms of visual appearance of the decoded images the situation is somewhat more involved. To illustrate the trade-off between ringing noise and blocking artifacts, we show coding results for two image coders employing filter banks with very different characteristics. Both filter banks are based on two-band allpass filters, used in a 3-level tree structure to produce a total of 8 subbands in each image dimension.

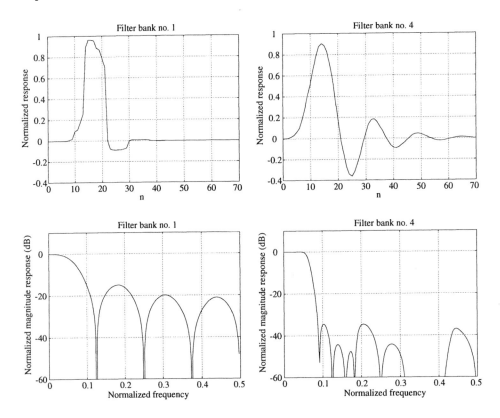

Figure 8.22: *Time and frequency characteristics of IIR filter bank lowpass channels.*

In Figure 8.22 both time and frequency responses of the two filter banks used are shown. The second filter bank (filter bank no. 4^2) is a "decent" filter bank in terms of stop band attenuation and channel separation. The other (filter bank no. 1), however, would be judged as quite poor from a classical point of view.

---

[2]The reasons for this numbering will become apparent later.

Figure 8.23: *Extract of original "Boat" image.*

In Figure 8.23 we show an extract of the image "Boat". The complete "Boat" image is coded at 0.35 bits per pel using the two aforementioned filter banks. Figure 8.24 shows the same extract as in Figure 8.23 of the decoded images obtained using filter banks no. 1 and 4, respectively. The quantization scheme is the same as in the previous section.

From Figure 8.24 it is clearly observed that, when using filter bank no. 4 we get severe ringing artifacts, whereas when using filter bank no. 1 these are avoided, but at the expense of blocking artifacts. This is not surprising since filter bank no. 4 is a typical representative of filter banks used in classical image subband coding. Filter bank no. 1, on the other hand, has a very limited region of support and behaves almost as a square transform. We also point out that in smooth areas, at some distance from edges and textures, filter bank no. 4 gives the most pleasing results. In areas with sharp discontinuities filter bank no. 1 performs better.

Based on these observations we have designed four different filter banks, each with different effective impulse responses lengths. The two filter banks mentioned above represent the extremes. In addition two other filter banks,

Figure 8.24: *Coding result using filter bank no. 1 (top) and no. 4 (bottom) at 0.35 bits per pel.*

no. 2 and 3, have been designed to have responses somewhat in between the two extremes. The filter coefficients of the allpass branches are listed in Table 8.3.

Filter bank no.	1	2	3	4
a_0	0.0000	0.0000	0.0000	0.0625
a_1	0.1000	0.3000	0.3500	0.5000

Table 8.3: *Filter coefficients for the allpass branches used in the adaptive filtering scheme.*

Blockwise Filter Switching

With four different filter banks to choose from, the idea is to switch between them according to the local image characteristics. In principle, we can allow for a steering signal $a_i(l)$ which changes value for every space index l, but this leads to a large bit rate overhead since the steering signal must also be transmitted. For reducing the side information, and to simplify the adaptive filtering scheme, we restrict ourselves to a blockwise filter adaption.

The original image is partitioned into 32×32 blocks, and based on the degree of edge content and dominant direction (horizontal/vertical), all blocks were classified into four groups and decomposed with one of these four filter banks. The classification is depicted in Figure 8.25 along with the original "Boat" image. In choosing filters for each block we simply inspected the coding result in the same block when using fixed filter banks 1 to 4. The filter bank giving the visually most pleasing result was chosen in each case – the only exception was when a block belonging to class no. 1 was adjacent to a block belonging to class no. 4. In this case filter bank no. 1 is used for the latter block also. This is necessary because the filter memory prohibits a sudden change of filter lengths.

When implementing the blockwise filter switching it is important that the signal decomposition is performed such that the switching takes place at the block borders. Figure 8.26 depicts the causal lowpass responses for all 4 filters, when using one filter stage only. In all plots the peak occurs at time lag 2.

```
(11) (41) (41) (41) (41) (41) (41) (41) (11) (11) (31) (41) (41) (41) (41) (11)
(14) (44) (44) (12) (44) (44) (11) (11) (11) (11) (31) (41) (41) (44) (22) (11)
(14) (44) (44) (12) (44) (44) (12) (11) (11) (31) (31) (21) (11) (11) (11) (11)
(14) (34) (14) (12) (14) (14) (12) (11) (11) (33) (44) (11) (22) (22) (22) (12)
(14) (34) (14) (13) (14) (34) (12) (11) (11) (24) (11) (11) (21) (22) (22) (12)
(14) (44) (22) (11) (11) (23) (12) (22) (12) (11) (11) (11) (11) (21) (21) (11)
(14) (44) (22) (11) (11) (11) (11) (22) (11) (11) (11) (11) (11) (11) (11) (12)
(12) (11) (22) (12) (11) (11) (12) (13) (13) (13) (11) (11) (11) (11) (11) (12)
(11) (11) (12) (22) (11) (11) (11) (11) (11) (11) (11) (11) (11) (22) (22) (12)
(11) (22) (22) (33) (22) (11) (21) (21) (21) (31) (21) (11) (31) (22) (22) (12)
(12) (22) (22) (22) (22) (22) (21) (21) (11) (31) (21) (21) (31) (22) (21) (12)
(11) (11) (11) (11) (11) (22) (33) (22) (33) (44) (22) (22) (22) (22) (12) (12)
(11) (31) (12) (22) (12) (33) (43) (44) (44) (44) (24) (22) (22) (22) (44) (14)
(12) (22) (22) (22) (22) (44) (44) (44) (44) (44) (23) (22) (22) (22) (42) (14)
(13) (43) (43) (43) (43) (43) (43) (42) (42) (42) (42) (42) (32) (22) (42) (12)
(11) (41) (41) (41) (41) (41) (41) (41) (41) (41) (41) (41) (41) (41) (41) (11)
```

Figure 8.25: *Original "Boat" image with block partitioning indicated (top), and classification of 32×32 image blocks (bottom). (ij) indicates that filter bank no. i is to be used horizontally and filter bank no. j vertically.*

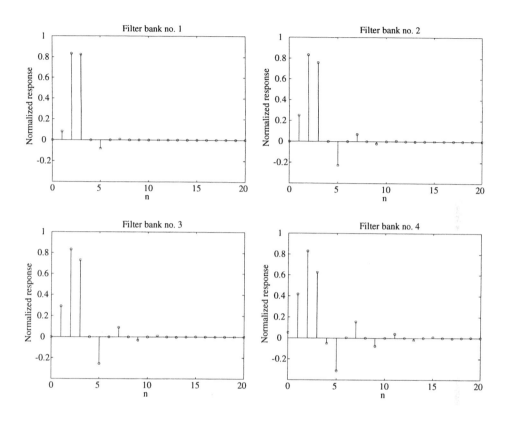

Figure 8.26: *2-channel lowpass responses of IIR filters based on two first-order allpass filters.*

To achieve a reasonable correspondence between the block partitioning and the effective unit pulse responses, we must advance the input signal 2 pels prior to analysis filtering. This corresponds to having the peak value at time lag 0. In the synthesis filtering the upsampled subband signals are shifted back 2 pels before filtering, thus achieving zero time lag in the total filtering procedure.

In Figure 8.27 we show an extract of the decoded image at 0.35 bits per pel using the adaptive filter bank scheme. Comparing with the decoded images of Figure 8.24, we note that by using the adaptive filter bank the ringing distortion of standard subband coding as well as the block distortions of coders based on block transforms are reduced. Figure 8.27 also shows a coding result using the nonadaptive FIR parallel filter bank 32II_opt from Section 8.2. Clearly, the adaptive IIR filtering scheme results in more ringing, as well as blocking-type noise.

The side information involved in the presented adaptive scheme consists of information indicating to the decoder which combination of vertical and horizontal filters to use for a given block. For each 32×32 block this requires 4 bits, corresponding to an additional bit rate of 0.004 bits per pel. This is a small price to pay.

Aspects of Circular Convolution

In classifying the image blocks, we implicitly accounted for potential problems with the circular extension method. This is evident from Figure 8.25 where we always use filter no. 1 at the endpoints of each line or column to be filtered. Using filter no. 1, which is close to a pure transform, one avoids extensive ringing noise at image boundaries. In [102] it was demonstrated that the performance of a traditional, fixed IIR filter bank could be improved by switching to filter bank no. 1 at the image boundaries. Filter bank no. 2 was used in the vertical and horizontal direction in the image center. The first and last 16 rows (columns) were filtered with filter bank no. 1 in the vertical (horizontal) direction. Using this scheme the ringing noise due to circular extension was reduced to the level obtained when using filter bank no. 1 for the whole image. Note that this scheme does not require side information. It was reported that a border zone less than 16 pels gave considerably more ringing noise. This is due to the filter memory, which forbids abrupt changes in the filter length.

The concept of filter switching solves to some degree the ringing problems of circular convolution when using nonlinear phase IIR filters. As we saw in Section 2.4, linear phase FIR filters offer a very elegant solution: Mirror extension [15].

Automatic Filter Selection

In the previous section we determined which filters to use in different image blocks by visual inspection. This can be considered as a feasibility study to determine the potential of adaptive SBC. A complete coding system must include automatic analysis for filter selection. A complete SBC system is shown in Figure 8.28.

Figure 8.27: *Decoded images at 0.35 bits per pel: Adaptive filter bank (top) and 32II_opt (bottom).*

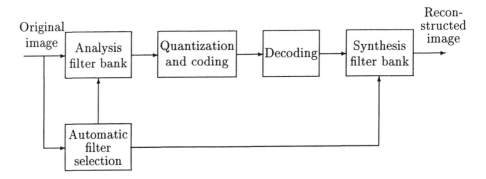

Figure 8.28: *Adaptive filtering with automatic filter selection.*

The task of automatic selection of filters for each image block is not trivial. In [102] the problem of filter selection was somewhat simplified by allowing filter no. 1 and 3 only, in conjunction with a block size of 16×16 pels. In the following subsection we give a brief overview of the method applied. For more details see [102]. This issue is also treated in [103].

Block Classification

Figure 8.29 shows the block classification scheme, in which the original image is partitioned into nonoverlapping blocks that are independently analyzed using edge detection techniques.

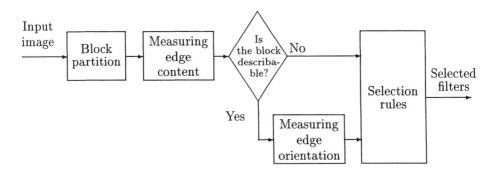

Figure 8.29: *Block classification for filter selection.*

By measuring the edge content in each block, it is classified into one out of three categories:

1. *Smooth*: The number of edge pels are so small that the block is assumed to be smooth.

2. *Describable*: The block has oriented structures that can be detected using the *Hough* transform [104].

3. *Too-busy*: The block contains too many edge pels. It is assumed to be heavily textured with no particular orientation.

In [102], all blocks with less than 4% edge pels were put in class 1, and blocks with more than 25% were put in class 3. The other blocks were put into class 2 where the Hough transform was used to find information about the edge orientation.

The final choice of filters is done in a separable manner using ad-hoc rules designed for visual optimization. They are summarized as follows:

- For contiguous blocks, which are either: 1) *smooth*, 2) *describable* with edges having the same orientation as the filter direction, or 3) *too-busy*, filter bank no. 3 is used.

- For contiguous blocks which are *describable* with edges having orientation normal to the filter direction, filter bank no. 1 is used.

- After initial classification, some adjustment is performed: If adjacent blocks are to be filtered with different filter banks, the block destined for filter bank no. 3 is reclassified to be filtered with filter bank no. 1. This is necessary because the filter response length should not change abruptly when switching. This is due to the filter memory.

Figure 8.30 shows the coding results obtained when using automatic filter selection. The bit rates are 0.35 and 0.26 bits per pel for "Boat" and "Lenna", respectively. For "Boat" the result is almost as good as for the manually classified image (Figure 8.27 (top)), whereas the "Lenna" image suffers from blocking effects. This is an inevitable result from our classification scheme, since we have made a strong emphasis of choosing filter bank no. 1 in the vicinity of any edges.

Figure 8.30: *Decoded images using automatic filter selection in an adaptive IIR filter bank.*

8.4 Dynamic Entropy Coding

In this chapter we have so far basically looked at the influence of the filter bank (basis functions) on the coding quality using fixed quantizers/coding strategies. The chosen digital representation for the subband components has been the JPEG-like scheme incorporating component thresholding, zig-zag scanning of the resulting component values in the different bands, run-length coding of the obtained string, and finally Huffman coding of the number pair representing a run and the next nonzero amplitude. This is the basic technique employed in JPEG and the existing video coding standards (H.261, MPEG1, MPEG2). We have chosen this scheme to be able to compare results. In the following we want to scrutinize this method in light of the quantization theory presented in Chapter 4.

It is important to note that if we have succeeded in the interband decomposition – as is the case with ideal filters and optimal transforms, and a very good approximation for practical filter banks – the components belonging to the same location (which will be part of one zig-zag scan) are uncorrelated. This remains to be true also after thresholding and quantization. However, the different components have different statistics. In this section we describe a method for efficient entropy coding of such signals appearing in subband coders.

If the zig-zag string were stationary, it would, according to the theory in Chapter 4, suffice to use one fixed entropy coder for the complete string. However, due to the fall-off of the input signal spectrum and the nonstationaries of real-world images, we have a mixture source with nonstationary parameters, indicating that an *infinite* number of entropy coders is necessary to obtain minimum rate for a given quantization. In that case we are left with an unbearable overhead bit rate necessary for informing the decoder about which entropy coder has been used for a given component. Therefore, we have to device a simpler scheme.

Let us for a moment look at how nonstationarities are coped with in systems employing *bit allocation*. The dynamic bit allocation approach allows each *block* of the image components (see Section 6.2.2) to use a number of bits computed from the bit allocation formula (Equation 4.42). There are usually a finite number of quantizers available, one for each integer number of bits. In a practical coder this number could typically be 7, representing from 0 to 6 bits.

In Chapter 4 we demonstrated the entropy-increase in a coding system

as a function of the allowed number of entropy coders when the signal is
nonstationary in terms of the variance. We concluded that for the special
example presented, the use of only one entropy coder, gives a loss of approximately 0.47 bits/pel compared to an infinite number of coders at high rate.
Most of this discrepancy is removed by using e.g. 7 entropy coders – the
same as the number of quantizers used in a typical bit allocation system.

We now present a practical coding experiment that tries to exploit the
advantage of using more than one entropy coder [105, 106]. We will also
compare the results with the JPEG-like method used earlier in this chapter.
Experiments are necessary for assessing the difference between the various
systems, partly because the image statistics do not necessarily obey any
simple statistical model, but also because a thorough theoretical evaluation
of the method combining run-length and entropy coding does not seem easy
or even feasible. The coding rate will throughout this section be estimated by
the first order entropy for all compared coding schemes. The actual rates will
be slightly higher due to nonexistence of perfect entropy coders. In practice,
e.g. an adaptive, multi-letter alphabet arithmetic coder will increase the rate
by only a few percent as compared to the theoretical first order entropy. We
account for this by adding 7% to the rate of all coders.

We will in this section use *filter bank 16I* and *the same uniform quantizer*
with equal thresholding ($t = \Delta$) for all schemes. The only difference will be
the entropy coding strategies. As the quantization is always the same, the
appearance at a given SNR will, of course, always be identical.

We compare three different entropy coding schemes:

1. The first coder employs adaptive entropy coder allocation. The quantized subband samples are grouped in blocks of 4×4 pixels and the rms
   value of each block is calculated and used for estimating the block's
   entropy. Each block is thereafter allocated the best matching entropy
   coder. Five different entropy coders are used. This number is a trade-
   off between the entropy gain and the amount of side information. A
   larger number of entropy coders would require more side information
   (indices pointing to the correct entropy coder). The low-low band is
   not included in this scheme, but rather represented by 5 bits DPCM
   coding.

2. The second coder uses only one entropy coder for all subbands, thus
   avoiding side information, though at a loss in bit rate due to less adaptivity than the coder using more entropy coders. The rate estimate is

based on entropy indicating that the optimal entropy coder has to be used, thus requiring initial transmission of the entropy coder parameters (e.g. the Huffman table). This extra rate is not accounted for in the plots below.

3. The third coder model applies JPEG-like coding, like the previously presented coders in this chapter. Also her the entropy coder parameters need to be transmitted as side information.

Figure 8.31 shows coding results for the 512 × 512 pixel image "Lenna".

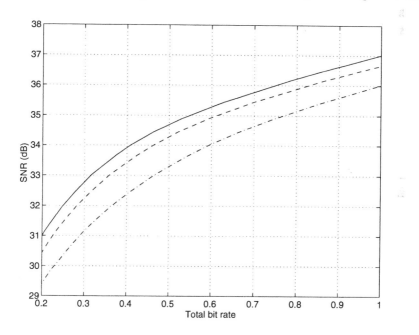

Figure 8.31: *Comparison of gains for different entropy coding strategies. Results: Curves from top and down: Coder with entropy coder allocation and efficient representation of side information (solid line). Subband coder with JPEG-like coding (dashed line). Coder with one entropy coder (dotted-dashed line). Image: "Lenna", green component.*

The gain increase when going from one to five coders is substantial. The coder using JPEG-like coding is better than the subband coder with one entropy coder, because the ordering of the subband samples in combination with the run-length coding exploits the characteristic zero-regions in the zig-zag string. The performance gain over the single entropy coder scheme is due to the correct entropy coder allocation. It is emphasized that the rate for the new coder scheme is the *total* rate, i.e. the first order entropy of the side information is included.

The main objective of this section has been to show how correct entropy coding can be used in subband coding. The conclusion is that different entropy coders should be allocated dynamically to the different channels based on the estimated entropy. The theoretical result found in Chapter 4 has been verified in the sense that a substantial rate decrease is obtained by increasing the number of entropy coders from one to five. The rate difference is approximately 0.25 bits/pel at SNR=36 dB. This should be compared to the high-rate theory which predicts 0.41 bits/pel (see Figure 4.11). Accounting for the transmission of side information (approximately 0.04 bits/pel), the correct number for comparison is 0.37 bits/pel. The correspondence is even better at higher rates. The merits of the JPEG-like entropy coding must be attributed to the ability of the run-length coder to exploit the inherent structure of the zig-zag scans.

We do not claim that the coder presented here is the ultimate coder for combining rate allocation and entropy coding. The ideas presented in this section are well suited for coders intended for image storage or transmission over data lines. The coding strategy is not limited to subband coders, but would also improve transform coders.

Although not treated here, the proposed model is also expected to perform better than bit allocation based methods. This is due to the inferiority of such schemes unveiled in the high-rate theory of Chapter 4. The inferiority of practical bit allocation methods also stems from the difficulty in using symmetric mid-tread quantizers at low rate, which is essential for visually successful representation of smooth image areas at low rates.

8.5   Summary

The aim of this chapter was to derive methods for improving the visual appearance of subband coded images at low bit rates. We have emphasized the influence of the filter bank on the distortion characteristics, rather than postulating the filter bank and subsequently designing suitable quantization schemes.

In Section 8.2 we have designed parallel, uniform, FIR filter banks to be used at low bit rates. A very general design procedure, described in [20, 21], has been adapted for this purpose. In particular, we have designed filter bank systems where the synthesis filter responses have lengths different from those of the analysis filters, and where the length decreases with increasing frequency. This was done for reducing the ringing noise associated with traditional image subband coders. In this design scheme arbitrary error criteria can be included in the optimization. We have mainly been concerned with the following aspects:

1. Perfect reconstruction error.

2. Coding gain optimization for an AR(1), $\rho = 0.95$ stochastic process.

3. Blocking effects error reduction, defined in Section 8.1.1.

The obtained filter banks seem to represent a good compromise between coding gain, blocking-free reconstruction of smooth image areas at low bit rates, and ringing noise close to edges.

The obtained filter banks have been studied in the space and frequency domains, and compared to the LOT and the DCT. It was possible to identify the important filter bank properties influencing coding gain and blocking artifacts. This confirmed that the extra degrees of freedom offered by nonunitary filter banks improve the coding quality.

In Section 8.3 we have established that IIR filter banks based on allpass building blocks can be made spatially adaptive without compromising their perfect reconstruction properties. By switching between 4 such filter banks with quite different characteristics in terms of the effective length of the impulse responses, decoded images with a low degree of ringing noise in image parts with sharp edges are obtained. At the same time blocking artifacts are not a problem. We might claim that the good features of transform coding as well as subband coding are combined in one single coding scheme, while avoiding the disadvantages of both.

The filter bank switching between blocks was first done manually, subsequently a simple, automatic classification algorithm was introduced. The results indicate that a good trade-off between ringing noise and blocking effects can only be made when the filter selection is good. The automatic filter selection in the presented example did not prove sufficient due to its tendency to produce blocking effects.

A comparison between an optimized, parallel, nonadaptive FIR filter bank with a tree-structured, adaptive IIR filter bank indicated that the adaptive system is less robust with respect to ringing noise. This is attributed to the fact that the channel unit pulse responses cannot change abruptly due to the filter memory. A better way of fast adaption would be an advantage in this respect.

The second main topic of this chapter has dealt with entropy coding methods for the quantized subband signals, in view of the theory in Chapter 4. It was demonstrated through a practical example that entropy coder allocation is a viable technique which is more efficient than the JPEG-like method used elsewhere in the chapter. This section also verified that there is a substantial gain in using more than one entropy coder for nonstationary sources. The results presented are not the best possible for the suggested model. Other choices of threshold values are e.g. better, but we have maintained the same value $t = \Delta$ to make fair comparisons between methods. Also variable rate coding of the DPCM prediction error would improve the results [66].

Chapter 9

Subband Coding of Video

9.1 Introduction

In the previous chapters we have seen that *image data compression* is obtained as a result of the removal of spatial domain redundancies and irrelevancies. Digital video, or digital image sequences, can be viewed as a discrete time sequence of digital still images or as a three-dimensional signal. Consequently we now have an additional dimension along which dependencies can be exploited for the purpose of bit rate reduction.

The importance of digital video compression is probably greater than that of still image compression because of the large existing base of services and applications involving moving images. Also, the successful solution to the problem of bit efficient representation of digitized moving imagery at a reasonable level of quality is recognized as being a key factor in paving the way for novel services and applications. The huge commercial potential of data compression for digital video is appreciated in the research community as well as in industry. This is the reason for the tremendous level of activity in this area as is evidenced by the steady increase in products as well as scientific publications.

The present chapter aims at giving the reader an overview of data compression of digital image sequences from a *subband coding* point of view. Although we focus on subband coding in the "classical" sense, we emphasize that techniques such as transform, wavelet and pyramid coding are particular instantiations of a subband coding scheme.

The chapter is organized as follows: In Section 9.2 we give a brief historical introduction to image sequence coding in general through the description of conditional replenishment coding. Subsequently, the major generic approaches to coding of moving images are described. This includes *hybrid coding*, i.e. a combination of two or more bit reduction techniques. Since many of these schemes make use of *motion compensation*, the topic of motion estimation from digital image sequences will be treated in some detail. Having described some generic coding structures, we proceed to present schemes employing *subband decomposition* while fitting into the general framework of the described generic structures.

The scientific papers on video subband coding can be classified into several categories, depending on which distinguishing features are emphasized. In this context we will split the presentation into two major parts, depending on the targeted bit rate of the various coding schemes. Based on this we describe low-to-medium rate coders (64 kbps – 5 Mbps) in Section 9.3, and medium-to-high rate coders (above about 5 Mbps) in Section 9.4. In the latter section an exposition of some of the techniques for subband coding of digital High Definition TV (HDTV) signals is given.

A word of warning is appropriate at this point: In this chapter a multitude of video subband coding approaches are presented. The authors have experience with simulation models of some of the techniques presented, but far from all. This fact, and other difficulties associated with comparing algorithms, make any attempt at picking the "best approach" dubious. Our attitude has been to present selections of what we believe to be important contributions to the field by emphasizing the features of the various approaches, but being careful in assessing relative performances. Finally, in presenting the various coding techniques we tacitly assume monochrome images to simplify the explanations. The extension to color images is straightforward as shown in Section 6.3.

9.2   Approaches to Coding of Digital Video

The simplest way, both conceptually and in terms of implementation, of achieving data compression for digital image sequences, is to treat each image frame as an isolated still image and apply some standard still image compression technique. This approach is referred to as *intra-frame coding*.

Although this approach is envisaged in some high-end applications[1], in this section we concentrate on *inter-frame* techniques in which the dependencies between subsequent image frames are exploited. In the following we shall treat:

- Conditional replenishment coding.

- Hybrid coding.

- Coding with motion compensation.

- Three-dimensional coding.

9.2.1 Conditional Replenishment Coding

Frame replenishment coding of moving images has proven successful in some applications. As an example we mention that video-conferencing services based on frame replenishment coding at 3 Mbps has been in commercial operation [107].

At present, frame replenishment coding is a well established and understood technique. Research in coding of moving images are therefore taking other directions. The technique is, however, important both in its own right as well as a basis for more advanced coding methods.

Early Schemes

The underlying assumption of conditional replenishment coding of moving images is that there is substantial correlation between picture elements in successive frames. This is particularly true in video-telephone and video-conference settings where the camera is stationary. In such scenes it is generally true that only a small part of the picture is changed from one frame to the next. For this reason one only needs to transmit information associated with the changed areas of the picture.

In his 1969 paper Mounts [108] describes a basic conditional replenishment coder. He is coding a television scene of a "head and shoulder" view of a single person engaged in conversation. The scene is recorded with 60 frames per second. He observes that in this particular scene, which he regards as typical, less than 10% of the total picture area is changed by more

---

[1]See Section 9.4 of the present chapter.

than 1% from frame to frame. In this coding scheme such a 1% change is regarded as significant.

The coder structure employs a frame memory to store the previously decoded frame as a reference. If a pel in the next frame to be coded is different from the corresponding pel value in the reference frame memory by more than 1%, then the pel value along with its address is sent to the receiver. In the original scheme this is done by 8 bit PCM coding of the pel's gray value, and by specifying a 7 bit address for each such pel relative to the start of a scan line. Thus, for each changed picture element a 15 bit word is transmitted. This basic conditional replenishment coder structure is depicted in Figure 9.1 a).

It is obvious that the size of the area of changed pels will depend on the motion in the scene to be coded. A consequence of this is that the bit rate will be time-varying. Therefore, if the coded image sequence is to be transmitted over a fixed rate channel, a buffering scheme must be implemented. If a buffering scheme is implemented, it is common practice to have the buffer fullness control the bit rate generated by the coder through the use of a feedback scheme. In [108] this was done by increasing the "significant-change-threshold" as the buffer filled up. Under all circumstances, given the finite probability of buffer overflow, any practical coder must include some mechanism for dealing with this situation [83].

The main results reported for this scheme can be summarized as follows:

- With the described coding scheme using on the average 1 bit per pel, the picture quality is nearly the same as for 8 bit PCM coding.

- If a buffering scheme is implemented, it is observed that as the motion in the scene increases the buffer fills, which in turn results in an increase in the "significant-change-threshold". Thus, the decoded image frame is not sufficiently updated when there are small, but perceptually significant, changes in the incoming frames. The resulting distortion, commonly referred to as the "dirty window effect", manifests itself as old image content being stuck in the decoded image frame.

We remark that although considerable attention has been paid to buffering schemes to allow transmission over fixed rate channels, attention is now also focused on variable bit rate coding for example in conjunction with packet transmission of digital video [109] and digital storage applications.

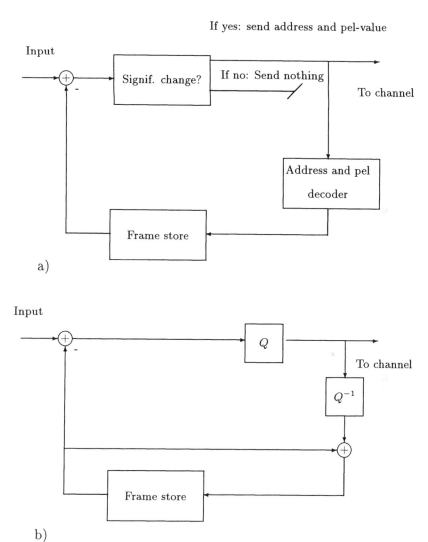

Figure 9.1: *a) Basic conditional replenishment coder conforming to [108].* *b) Simplified block diagram of improved conditional replenishment coder employing predictive coding of changed areas. The block labeled Q performs the quantization of the prediction error signal. This basic structure forms the core of most hybrid video coding schemes.*

Improved Conditional Replenishment Coding

Subsequent to Mounts' contribution [108], numerous modifications and improvements to the basic scheme have been proposed and evaluated. The major enhancements can be categorized as follows:

- Predictive coding of changed areas.

- Variable word length coding of updated pels (entropy coding).

- Improved addressing strategies.

- Improved segmentation of the image.

- Temporal filtering.

- Subsampling in rapidly moving areas.

Here we shall only comment on predictive coding. Details on the other items above can be found in [110].

In Mounts' original scheme each pel in the moving area was coded independently by PCM. It is possible, however, to exploit temporal as well as spatial redundancy in the moving area by predictive coding. Thus, rather than sending the quantized value of a pel, the quantized value of the prediction error is sent. This is depicted in Figure 9.1 b). It is well known that the efficiency of a predictive coder increases as the variance of the prediction error, relative to the variance of the signal that is predicted, decreases. A reasonable measure of performance for a specific predictor should therefore be the root mean square (rms) value of the prediction error as found by applying the predictor to several typical image sequences. In many cases the entropy of the (quantized) prediction error has also been used as a measure of performance. This is a very meaningful measure due to the experience that one can design entropy-codes performing very close to the computed entropy rate [111].

In [112] several causal inter-frame, intra-frame and inter-intra-frame predictors where studied. It was found that for low motion scenes, inter-frame predictors are more efficient than intra-frame predictors. As the motion in the scene increases, intra-frame prediction becomes more competitive. This is in part due to the fact that spatial correlation between adjacent pels increases with motion due to the integration inherent in the recording camera [112]. Also, the frame-to-frame correlation is higher the less motion is present. In this context we point out that in areas of rapid motion, particularly if the motion is irregular and thus difficult for the human eye to track, the *required* spatial resolution is considerably lower than in stationary or slowly moving areas. This has been exploited in schemes employing adaptive spatial subsampling [110].

Several more advanced predictive schemes than those described above have been reported and examined in the literature. These are primarily adaptive schemes that aim at optimizing the predictor coefficients with respect to the local statistics of the images [112, 113]. The schemes described in these references employ standard Wiener-theory supplemented by some ad-hoc ideas.

Conditional replenishment coding was the first technique for coding of digital image sequences exploiting *temporal* dependencies. Viewing the technique primarily as a way of removing redundancies/irrelevancies in the temporal domain, we might, in generic terms, think of it as a DPCM coding scheme operating along the time axis.

9.2.2 Hybrid Coding

The term hybrid image coding refers to the combination of two distinct techniques for bit rate reduction. The rationale for the application of hybrid techniques is either improved performance or reduced complexity. An example of the latter in the context of still image coding is presented by Habibi in [114]. In that paper still images are compressed by first applying transform coding along the scan lines of the image. Subsequently, DPCM is applied vertically to the transform coefficients. This technique is advanced as an alternative to full two-dimensional transform coding because of lower computational complexity. In the present context hybrid coding removes temporal dependencies by one compression technique, and the spatial redundancies/irrelevancies another technique. Almost invariably this involves DPCM in the temporal domain and some other technique such as transform or subband coding in the spatial domain. Limiting ourselves to this scenario two coding structures will be discussed:

- Inside loop subband/transform coding: Spatial subband/transform coding is done inside the temporal prediction loop on the prediction error. Sometimes this structure will be referred to as *model 1* [115].

- Outside loop subband/transform coding: The subband/transform decomposition is applied to each frame of the sequence to be coded. A predictive coder is applied to the subband/transform coefficients of subsequent image frames. Sometimes this structure will be referred to as *model 2* [115].

These two basic hybrid coder structures are depicted in Figure 9.2. In this figure and in subsequent figures we use the acronym FB for an *analysis filter bank* and FB^{-1} for a *synthesis filter bank*. Almost all current hybrid video coding schemes conform to either one of these categories. While the former is by far the most popular in transform based coders, the latter is receiving increased attention.

9.2.3   Motion Compensated Coding

In Section 9.2.1 was stated that as the motion in the scene increases, the effectiveness of the inter-frame prediction decreases. This is because the pel to be predicted has moved from a position in the previous frame that lies outside the region of support of the predictor. However, if one knew the *optical flow*, i.e. the displacement of each pel from the previous to the present frame, one could base the prediction on the corresponding pel-position of the previous frame *compensated for by the known displacement*. An illustration is given in Figure 9.3. This is the basic idea behind motion – or displacement compensated predictive coding. This technique has proven to be a powerful coding algorithm for a wide range of bit rates. If we allow noninteger values for the displacement we can compute the prediction of the luminance of a pel in the current frame by some sort of interpolation applied to pels in the previous frame. In practice, *bilinear interpolation* is often applied. When this is done the prediction is computed by Equation 9.1 below. See also Figure 9.4.

$$
\begin{aligned}
\hat{I}(\underline{x}, t) \;=\; & (1 - \alpha)(1 - \beta)\tilde{I}(x - \lfloor d_x \rfloor, y - \lfloor d_y \rfloor, t - \tau) \\
& + (1 - \alpha)\beta\tilde{I}(x - \lfloor d_x \rfloor + 1, y - \lfloor d_y \rfloor, t - \tau) \\
& + \alpha(1 - \beta)\tilde{I}(x - \lfloor d_x \rfloor, y - \lfloor d_y \rfloor + 1, t - \tau) \\
& + \alpha\beta\tilde{I}(x - \lfloor d_x \rfloor + 1, y - \lfloor d_y \rfloor + 1, t - \tau) \qquad (9.1)
\end{aligned}
$$

where

$\underline{x}$ is the 2-D position vector $[x, \quad y]$,

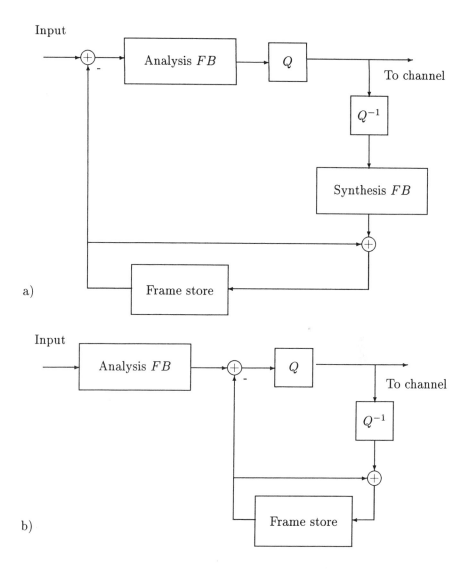

Figure 9.2: *Two hybrid image sequence coders: a) Inside loop decomposition (model 1), and b) outside loop decomposition (model 2).*

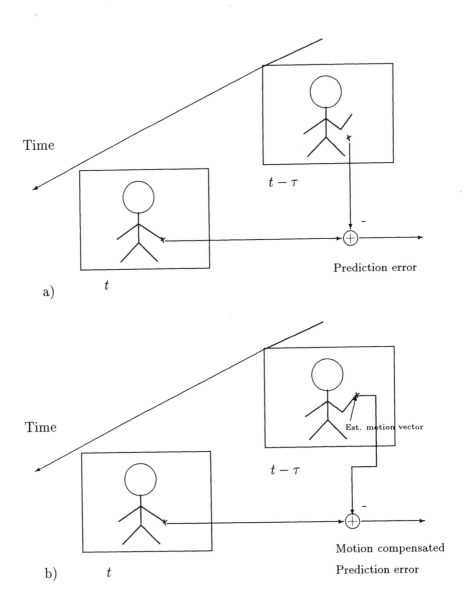

Figure 9.3: *Illustration of motion compensated prediction: a) Prediction based on corresponding pel of previous frame, b) Prediction based on pel at position compensated for by estimated displacement.*

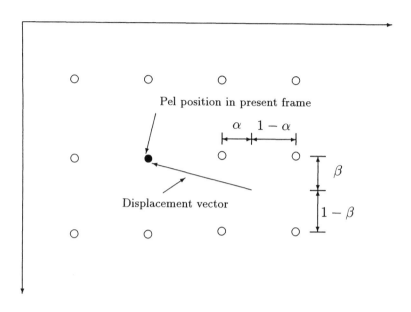

Figure 9.4: *The prediction for the pel in the present frame is based a bilinearly interpolated luminance value in the previous frame. The coefficients α and β are used in this computation, see Equation 9.1.*

$\hat{I}(\underline{x}, t)$ is the predicted luminance at position $\underline{x}$ and time t,

$\tilde{I}(\underline{x}, t - \tau)$ is the reconstructed luminance for the previous frame,

d_x, d_y are displacement estimates, of the pel in question, in the vertical and horizontal directions, respectively.

$\lfloor z \rfloor$ denotes the integer part of z.

The prediction error signal is then given by

$$e(\underline{x}, t) = I(\underline{x}, t) - \hat{I}(\underline{x}, t). \tag{9.2}$$

A central issue in coders based on motion compensation is the accuracy of the displacement estimate. In Figure 9.5 we have shown a block diagram of a motion compensated hybrid coder in which both the coded prediction errors and the motion estimates are transmitted to the decoder. The displacement compensation is, of course, helpful only in areas of the image sequence where there is motion.

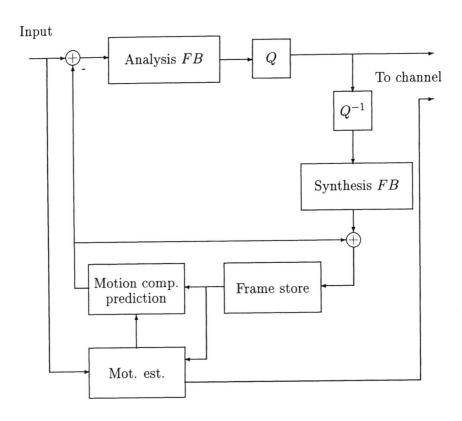

Figure 9.5: *Block diagram of motion compensated predictive coder with inside loop signal decomposition (model 1).*

In Figure 9.5 the remaining *spatial* redundancies in the prediction error image are removed by a still-image subband coder, to the extent that this is possible. This is the approach of both the H.261[86] and the MPEG [116] recommendations in which the motion compensated prediction error image is transform-coded in the spatial domain by a DCT coder tailored specifically to this type of signal. Several examples of schemes in which subband decomposition of the prediction error signal is employed will be presented in Sections 9.3 and 9.4.

Motion Estimation for Motion Compensated Coding

In pure predictive coders there are two dominant techniques for estimating the motion vectors needed in a motion compensated coding scheme, namely *pel-recursive* [117] schemes and approaches based on *block matching* [118]. The first technique has found little use in hybrid codecs. One exception is found in [119]. Block matching has, on the other hand, been employed extensively in hybrid coding schemes. Due to the importance of block matching in the context of hybrid motion compensated coders, we shall concentrate on this technique.

Block matching for motion estimation   In block matching the image is partitioned into rectangular blocks (typically of size 8 × 8 or 16 × 16 pels). Each block in the current frame is compared to all possible blocks in the previous frame[2], but limited to a search window of predetermined size. The spatial offset of the most similar block in the previous frame according to some similarity measure, is taken as the motion estimate for the whole block in the current frame. This technique, – when the most similar block is found through exhaustive search within the search window, is referred to as *full search block matching*, and is illustrated in Figure 9.6.

The number of distinct offsets to be evaluated in this strategy is given by $(2d+1)^2$ where d denotes the maximum horizontal or vertical displacements of the search. A typical value for the maximum displacement in low bit rate video coders is $d = 7$. Many researchers have considered this number, $(2d+1)^2$, as prohibitively high, and much effort has been spent on finding computationally more efficient algorithms [118, 120, 121, 122, 123]. Common to these efficient algorithms is the assumption that the similarity, viewed as a function of the offset value, has one global maximum and no local maxima. This is not always the case, and the consequence is that the simplified algorithms have suboptimal performance. The computational burden of block matching is proportional to the required number of operations in calculating the error measure. The two most commonly used measures are the mean square error (MSE) given by

$$MSE(\underline{d}) = \frac{1}{HV} \sum_{\underline{x} \in block} (I(\underline{x}, t) - I(\underline{x} - \underline{d}, t - \tau))^2, \qquad (9.3)$$

---

[2]This may be either the previously decoded frame or the previous original frame. In the sections to follow examples of both cases will be shown.

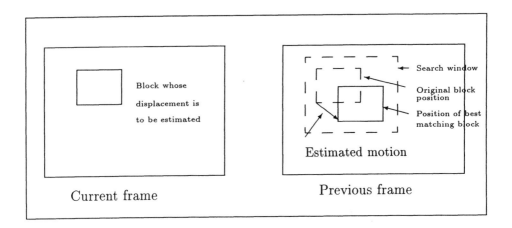

Figure 9.6: *Block matching for computing optical flow (displacement).*

and the mean absolute frame difference (MAD):

$$MAD(\underline{d}) = \frac{1}{HV} \sum_{\underline{x} \in block} |I(\underline{x}, t) - I(\underline{x} - \underline{d}, t - \tau)|, \qquad (9.4)$$

where H and V are horizontal and vertical dimensions of the block, respectively. From a hardware implementation point of view, the latter measure is simpler.

Given the advantage of hindsight, it seems that the benefits of the work on reduced complexity search strategies for block matching do not totally justify the efforts. This is because current trends in VLSI implementation of block matching favor the full search algorithm because of its *regular* structure [124].

We conclude this subsection by pointing out that block matching has gained widespread popularity due to its conceptual simplicity and usefulness in motion compensated coders. It is known, however, that it is rather inferior in rendering the *true motion* in a scene. In most image sequence coders the optical flow field, computed by means of block matching, is transmitted to the decoder. The motion estimates delivered to the decoder are represented exactly. More advanced methods for computing optical flow fields that are very close to the true motion have also been investigated [125]. In that study the superior motion estimates were coded with loss and transmitted to the decoder. For several bit rates it was found that to exploit the reduced prediction errors resulting from the superior motion estimates, an excessive number of bits had to be used for the representation of the motion vectors. Thus, the technique did not offer advantages over schemes based on block matching in a coding context. Using better motion estimates is still believed to hold some promise when incorporated in schemes for optimally distributing the bits between the prediction error signal and the motion estimates.

9.2.4 Three Dimensional Coding

Subband decomposition can obviously be extended beyond two dimensions. A digital image sequence can be viewed as a three dimensional signal that can be split into subbands both spatially and *temporally*. As in one- and two-dimensional subband coders, this facilitates the removal of redundancy in the signal. The three-dimensional subbands may also be viewed as signal entities of different perceptual importance and as such they may be assigned bits accordingly. Several examples of this generic scheme will be treated in a subsequent section of this chapter.

9.3 Low-to-medium Rate Coding

Research on image sequence coding at low to medium bit rates (i.e. below $\sim$ 5 Mbps) has to a large extent concentrated on hybrid coding schemes incorporating motion compensated prediction in the time domain and transform coding, employing the DCT, in the spatial domain. Although the DCT based schemes have been dominant, increasing interest has been directed towards alternative techniques, such as subband coding. There are disadvantages associated with transform-based schemes, the most noticeable being the *blocking effect*. This problem is to some extent circumvented in subband coding schemes, although other artifacts may occur in these schemes at low rates.

In this section we present several low-to-medium rate subband coding schemes. Most published techniques fall into the class of hybrid coders with subband decomposition *inside* the prediction loop [126, 127, 128, 129, 130,

131, 132, 133, 115] (i.e. model 1 structures). In most cases motion compensation is employed. Other researchers have applied subband decomposition to the original image frames and applied temporal prediction to the subband images [115, 134, 135, 93] (i.e. model 2 structures). We also provide some comments regarding the relative merits of the two approaches in light of recent research. Finally, techniques relying on full three-dimensional subband decomposition are explored [136, 137, 138, 139]. Within the confinements of these three main approaches, the reported schemes differ primarily in the types of filter banks employed and in the approach taken to the bit efficient representation of the subband signals.

9.3.1 Subband Decomposition Inside the Prediction Loop

The application of temporal prediction removes a substantial portion of the temporal correlations in an image sequence. Also, if the prediction works well, large portions of the prediction error image will have values close to zero. This is illustrated in Figure 9.7 with the original image frame no. 25 of the CCITT sequence "Miss America" along with prediction error images for the case when no motion compensation is applied, and for the case when motion compensation is applied[3]. In comparing the original frame to the corresponding prediction error images, we observe that some of the active parts of this image have a *highpass* characteristic. In these areas spatial correlation will be present. Also, there are portions in a prediction error image corresponding to newly uncovered background. These areas exhibit the same type of correlation properties as the original previous frame. Large areas of the prediction error image is very close to zero. These areas are easily represented by few bits. Although not evident in this particular example, there will also be portions where the prediction error image is spatially almost uncorrelated, but still of visual importance. Thus, when applying still image compression techniques to the prediction error image, large coding gains are not expected in all areas of the prediction error image. Based on these observations it seems plausible to employ three different strategies, or modes, in a bit efficient coder for the prediction error image [140]:

---

[3]The zero level has been adjusted to a medium gray level for display purposes.

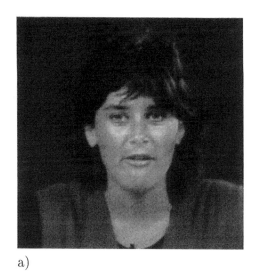

a)

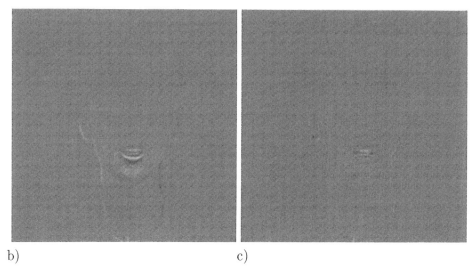

b)   c)

Figure 9.7: *Frame no. 25 of the "Miss America" sequence: a) Original, b) prediction error without motion compensation, c) prediction error with motion compensation. For display purposes the zero level has been translated to medium gray level value.*

1. Portions of the prediction error image with values close to zero need not be transmitted at all[4].

2. Areas exhibiting low sample to sample correlation, as mentioned above, should be represented bit efficiently in some way or another. In these areas one should not expect large coding gains. In [140] such areas were coded using what essentially corresponded to an adaptive scalar quantizer.

3. Parts having high correlation are compressed with significant gain using traditional still image coding techniques.

These statements should be contrasted with current practice in image sequence coding in which the complete prediction error image is decomposed using a transform or a filter bank. Based on the observations above one might argue as follows:

- For the uniform areas of the prediction error image, with gray level values close to zero, there is no need for any transformation or subband decomposition, since no information about these areas needs to be transmitted anyway. What exactly constitutes a level close to zero is, of course dependent on the targeted bit rate for the coder. If the prediction error signal is quantized with a "dead-zone" quantizer, see Figure 4.2, what is considered *close to zero* is determined by the selection of Δ.

- For the low correlation areas of the prediction error image, transform or a subband decomposition does not help much. Rather one could address the problem of efficiently representing relatively uncorrelated signals as a problem in its own right [140].

- For the parts resembling still images, traditional still image coding based on transforms or subband decompositions, should be applied because it is expected to work well.

In defense of the hybrid coding schemes presented earlier, the following remarks can be made:

---

[4]Of course, information indicating the location of these areas must be made available to the decoder.

- Concerning the uniform areas of the prediction error images that need not be transmitted, performing a transform or a subband decomposition *does not hurt*. Typically, if a sensible scheme for the quantization of the subband signals or transform coefficients is employed, these areas will still not produce significant information requiring the transmission of bits.

- Areas with low sample to sample correlation still do not suffer by being subjected to a signal decomposition.

- The portions in which there are substantial sample to sample correlation are efficiently coded by use of a subband or transform scheme in the same way as still images are.

From these deliberations we might attempt a conclusion: For the areas of the prediction error image requiring no bits and for those areas characterized by low sample to sample correlation, there is no gain in terms of compression to be expected by performing signal decomposition. In the areas where there is such correlation, gain is obtained. Therefore, in low activity scenes where the proportion of high correlation areas will be very low, the total gains by applying a transform or a subband splitting are modest. For more complicated, higher activity scenes, a larger portion of the prediction error image will stem from uncovered objects/background implying higher correlation in the prediction error image. In this situation the benefits of applying a subband scheme increase. Therefore, by using a hybrid scheme we will never lose anything, but the gains in terms of compression efficiency may not always be great. We should also keep in mind that *if* we do want to exploit the possibly small gains attributable to subband/transform decomposition of the prediction error image, we will get a *regular* structure for the coding algorithm if the same treatment is applied to the whole prediction error image. In the case of block oriented transforms we could conceivably obtain lower complexity by only applying the transform to areas where this would give rise to coding gains, and use simpler strategies when dealing with other parts of the prediction error image. In terms of hardware this might, however, lead to a more complex solution even though the number of arithmetic operations is reduced. The reason is that the algorithmic regularity when applying the transform to all parts of the prediction error image is lost. Also, the application of filter banks involved in a subband decomposition to specific areas of the prediction error image is not straight forward, since the unit pulse responses are overlapping. Therefore, from a pragmatic point of view, the hybrid coding approach with transform/subband decomposition inside the prediction loop does make sense in spite of some of the objections cited above. Acknowledging these issues, we now proceed by describing several important contributions to low and medium bit rate subband coding of digitized image sequences. The main differences between the various approaches lie in the filters selected, the representation of the subband signals and the number of subbands employed.

Video Subband Coding with FIR Filter Banks and Entropy Coding

A good representative of the model 1 scheme was proposed by Gharavi [133, 115, 134]. The structure of this coder, operating at about 64 kbps, is shown in Figure 9.8 a). The filters used in the filter bank were the 16 tap FIR filters referred to as f16a by Johnston [17]. These filters were applied in two-band structures forming the basis of a tree-structured filter bank giving rise to a 10 band spatial decomposition as depicted in Figure 9.8 b).

Unlike the still image case, for which the low pass band is given preferential treatment, all the subbands can be treated in essentially the same way with respect to quantization. Rather than applying a bit allocation strategy in combination with a pdf-optimized quantizer, it is argued that uniform quantizers with a "dead-zone" should be used. This seems to be in accordance with the general theory presented in Chapter 4. The characteristic of such a quantizer is shown in Figure 4.2. For each of the 10 subbands a quantizer of this type is used. The parameters of the quantizer (i.e. the "dead-zone" and the quantizer step size) are found experimentally through a series of subjective evaluations. The characteristic features of the image subbands have an orientation in accordance with the location of the subband in the spatial frequency domain. Thus, the dominant structures in the various subbands can broadly be classified as horizontal, vertical and diagonal. This observation forms the rationale for the encoding strategy applied to the subbands: Each subband is partitioned into nonoverlapping blocks. Subsequently the blocks are scanned according to either a zig-zag, a vertical or a horizontal scanning pattern as depicted in Figure 9.9. The selection of one of these scanning patterns is dependent on the location in the spatial frequency domain of the subband to be scanned. This scanning of subband signals give rise to one-dimensional strings of signal samples in which a large

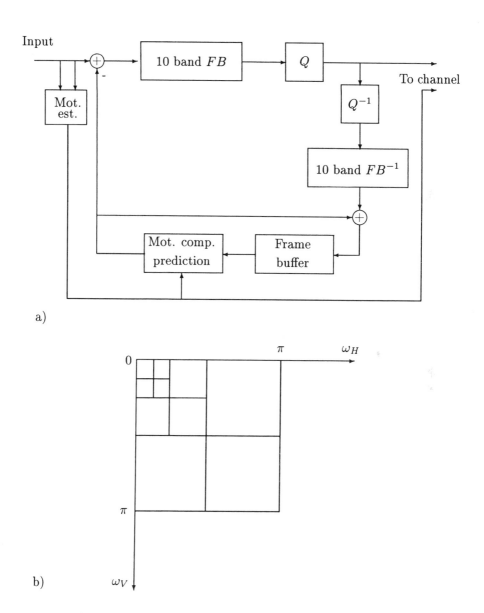

Figure 9.8: *a) Block diagram of subband video coder, b) 10 band spatial subband decomposition applied to the prediction error image.*

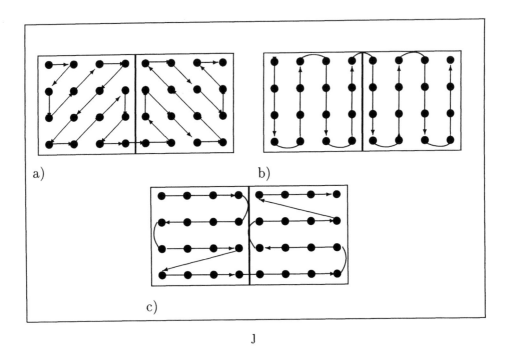

Figure 9.9: *Scanning patterns for blocks extracted from subbands having mostly a) diagonal, b) vertical, or c) horizontal structures.*

proportion will, depending on the activity of the subband block, be quantized to zero. Thus, applying *run-length* coding to these one-dimensional collections of quantized subband samples gives rise to substantial data compression. Finally, the run-length codes are compressed further by employing entropy coding. This results in variable length codewords (VLC). The experiments reported in [115, 134] indicate that this subband approach is preferable to some schemes based on the DCT at bit rates around 64 kbps.

Video Subband Coder with IIR Filter Banks

For a given magnitude response, an IIR filter that has a considerably more efficient realization than an FIR filter, can be designed. The practical significance of this in a still image subband coding context was demonstrated in [141]. In a real time video environment the issue of computational complexity is even more pressing. Therefore, IIR filter banks can be a viable alternative in the subband decomposition. Another solution is the use of FIR filters with short unit pulse responses, such as the *symmetric short kernel filters* (SSKF) [142] which were successfully employed in a video codec in [128, 129]. As an example of a low complexity scheme we shall give an outline of the video codec described in [131, 51]. A main feature of this subband coder is the extremely low computational complexity of the allpass filters employed in the filter bank portion of the codec. In this particular case a uniform 8×8 band decomposition, employing only first order allpass filters in a tree-structure, was used. The two-channel building block forming the core of this filter bank is described in detail in Section 3.4.3. The overall structure of the subband video coder incorporating motion compensation conforms to the generic structure of Figure 9.5.

Since the very low complexity of the IIR filter banks is emphasized, they are applied in conjunction with low complexity strategies for the bit efficient representation of the subband signals. In doing so, well known strategies for coding of DCT coefficients were adopted [85, 143, 86].

Coding of the subband signals. Given the strong relationship between subband and transform coding it is reasonable to expect that bit efficient strategies that have been successful in transform coders should also be useful in a subband coding context. Below such a technique based on the strategy of [86] is described.

Denoting prediction error subband no. (m, n) by $x_{m,n}(k, l)$, we can, for each spatial coordinate (k, l) within a subband, view $x_{m,n}(k, l)$ as functions of (m, n). We can then interpret $x_{m,n}(k, l)$, (k, l) fixed, as a set of "transform coefficients". For those areas of the prediction error image exhibiting sample to sample correlation, it seems plausible that the amplitude of these "coefficients" decreases as the frequency of the subband from which they are taken increases. Thus, scanning them in sequence of increasing spatial frequency, i.e. according to a zig-zag pattern as depicted in Figure 9.10, one would expect a length 64 vector with elements of decreasing amplitudes as the vector index increases. We refer to this 64-vector as a one-dimensional scan string.

Efficient representation of the scan strings is obtained by first thresholding, that is all samples with an amplitude below a certain value are set to

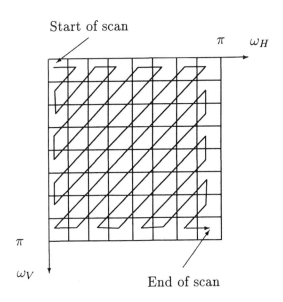

Figure 9.10: *Zig-zag scan of the subbands for an 8 × 8 uniform filter bank.*

zero. Subsequently a uniform quantizer is applied. Note that with a suitable selection of the threshold value, this corresponds to a uniform quantizer with a "dead-zone" around zero. Following this the one-dimensional strings of thresholded and quantized samples are represented by entities of the form $(level, run)$ where run designates the number of zeros preceding the sample of value equal to $level$. One scan string would then be represented as $(level_1, run_1), (level_2, run_2), \ldots, EOB$, where the EOB means that there are no more nonzero samples left in the string.

Example The string of thresholded and quantized signal samples

8 9 0 0 1 0 5 0 0 ... 0 would give rise to the coded sequence
(8,0), (9,0), (1,2), (5,1), (EOB)

Treating the above representation of a scan-string as source words, a single Variable Length Code (VLC) is designed. In evaluating subband coders based on this scheme entropy rates are quoted. Applying actual entropy codes gives rise to a bit rate that is only marginally higher than the computed entropies.

Motion compensation. In the low complexity video coder described here, block matching is employed for motion compensation. The block size was set to 16×16 pixels and the maximum displacement used was ± 7 pixels[5], both vertically and horizontally. Furthermore, based on current trends in VLSI for performing block matching [124], the block matching was performed with *full search*. The function minimized in the search was the *mean square error* between a block in the current frame and the displaced blocks of the previous decoded frame.

It is well known that straightforward application of a block matching algorithm may give rise to nonzero motion vectors even when there is no actual movement. For this reason it is common to apply a classification scheme to all the blocks either prior or subsequent to performing the matching. In the work presented here the approach of [144] in designating a block as "moved" or "nonmoved" was employed.

The method used for coding of the motion estimates was based on [145, 146, 147]. There seems to be consensus that the preferred way of coding the motion vectors, as obtained by block matching is:

- Code motion differentials rather than directly coding the motion estimates. That is, along each horizontal line of blocks we code the difference between horizontally adjacent motion estimates.

- Treat each *vector* of motion differentials as the source word to be coded, that is we code the vertical and horizontal displacement differentials together rather than separately as is done for example in [144].

Akansu [147] concludes that an adaptive Huffman code [148] is suitable for this purpose and that this method of coding performs very close to the entropy of the motion vector differentials. This is also the approach taken in the IIR based subband coder presented here.

Results. Simulations presented in [131] showed that the presented scheme had a comparable performance in terms of SNR to an H.261-like coder based

---

[5]These choices are the same as those in [144].

on the DCT at bit rates around 64 kbps. Subjectively there were minor differences due to less blocking. The IIR filter banks used were implementable with no multiplications leading to extremely low computational complexity. The codec was also compared to a similar one employing the 16B FIR filters of Johnston [17], see also Chapter 3. No visible differences in subjective quality from the IIR based coder were observed. Thus, the reduced complexity of this coder does not come at the expense of reduced quality images.

Before closing this section on the IIR based video coder we point out that the various modules of the presented coder are commonly used in many subband/transform schemes for coding of digital image sequences.

Conditional Replenishment Video Subband Coding

As a link to the next section on coders with subband decomposition outside of the prediction loop (model 2 structures), we close this section by presenting the 300 kbps video subband coder of Westerink [126].

Westerink, also used the generic structure of Figure 9.5 a), i.e. a hybrid coder incorporating motion compensation. A 13 band split, as shown in Figure 9.11, is applied to the possibly motion compensated prediction error image. The filter bank is tree-structured and based on the 16B filters of Johnston [17]. In contrast to the previously described strategies, giving rise to a time-varying bit rate, the present coder makes use of an explicit bit allocation algorithm, originally introduced in a still image coding context [69], so as to enforce a constant and predetermined bit rate for the coded image sequence. The subbands of the prediction error signal are divided into blocks of 4×4 samples. If the mean square value of these samples is below a preset threshold value, the block is not assigned any bits in the subsequent bit allocation stage. Note that this in essence is a conditional replenishment scheme, that is, if a block within a subband has not changed sufficiently the corresponding signal samples are not sent. This is indicated to the decoder by setting a single *conditional replenishment* bit (CR bit) for each 4×4 subband block to zero. If the threshold is exceeded, this bit is set to a one.

When this classification of subband blocks is completed, the subbands are assigned bits according to the bit allocation algorithm of [69]. At the targeted bit rate of 300 kbps, some of the blocks with activity levels above the threshold value of the block classification stage are located in subbands that have not received any bits in the bit allocation. The resulting distortion that is commonly referred to as the "dirty window" effect. This manifests

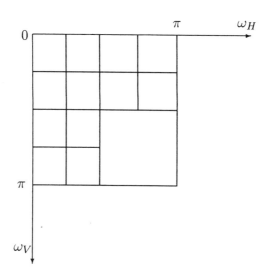

Figure 9.11: *The prediction error image is split into 13 subbands.*

itself as old image content being stuck in the decoded image frame. This happens because the image is not updated to reflect small, but still perceptually significant changes in the incoming frames. Among several possible remedies the one proposed in [126], called *zero block correction* (ZBC), can be explained as follows: Rather than not updating the subband block of the decoded frame (i.e. just copying the contents of the subband block of the same subband block in the previously decoded frame) the subband block in question is simply set to zero. Crude as this may seem, this is claimed to reduce the "dirty window" effect significantly. Unfortunately, in the coder structure presented up to this point, the subband decomposition of the reconstructed previous frame is not available. Therefore, to incorporate the ZBC the structure must be modified to the structure of Figure 9.12. We observe an additional analysis filter bank which adds considerably to the complexity of the scheme.

If the motion compensation, which is assumed to be in effect, is dropped it is easily shown that the coder can be simplified to the structure shown in Figure 9.13[6]. In [126] it is claimed that the benefits of motion compensation

---

[6]I.e. the two analysis filter banks are merged into one operating on the incoming

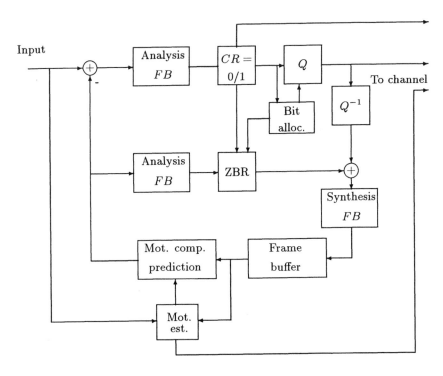

Figure 9.12: *Modified motion compensated coder structure incorporating zero block correction (ZBC).*

are so small, within the constraints of this particular coding technique, that they do not justify the additional computational complexity involved. In this context we would like to comment that this is somewhat in contrast to the more widely held belief that motion compensation is a most useful technique in low rate hybrid coders.

Having arrived at the structure of Figure 9.13, which in effect is a scheme in which the subband decomposition takes place *outside* the prediction loop, we conclude our exposition of low-medium rate subband coders with signal decomposition inside the prediction loop.

9.3.2   Subband Decomposition Outside the Prediction Loop

Although hybrid coders employing subband decomposition inside the prediction loop are by far the most popular at low to medium bit rates, several

---

frames and the single synthesis filter bank becomes unnecessary.

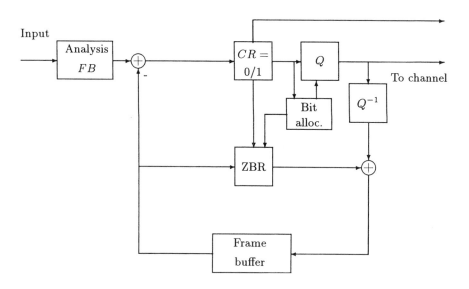

Figure 9.13: *Coder structure without motion compensation incorporating ZBC.*

researchers have investigated coding schemes in which the subband splitting is performed prior to forwarding the signals to the time domain prediction loop. In fact the very first papers on subband coding of moving images [149, 150] conformed to this structure.

Subband Coding with Hierarchical Motion Compensation

In his coding scheme, targeted at a bit rate around 320 kbps, von Brandt [150] employs a structure as shown in Figure 9.14. The subband splitting is performed using a tree structured filter bank. As basic building block a two-band structure, employing QMF filters with 7 or 13 taps, is used. The resulting 31 band decomposition is depicted in Figure 9.15. In performing the motion estimation needed for the motion compensated prediction, a hierarchical scheme utilizing low pass bands at intermediate levels of the filter bank tree is proposed. Specifically, the bands resulting from low pass filtering followed by a spatial subsampling by factors of 2 and 4, respectively, are used together with the original image frame in estimating the motion as follows: First, a coarse estimate of the motion is found by applying block matching to the low pass band subsampled by a factor of 4 in each direction.

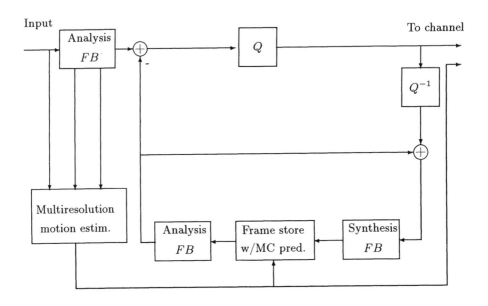

Figure 9.14: *Structure of the video subband coder [150].*

These coarse estimates, covering large displacements, are obtained at a low
cost in terms of the amount of computation. These estimates are used
as initial values in the next step of the estimation procedure where image
frames, decimated by a factor of two in each direction, are used. Given
the initial estimates obtained at the previous stage, the search area of the
block matching algorithm at this stage can be made smaller. In the final
stage, block matching, incorporating the initial estimates of the previous
stage, is performed with a small search area on the original image frames.
The advantage of this approach is that large displacements can be estimated
reliably at a reasonable computational cost.

    Since the motion estimates are obtained for the full resolution image
frames and *not* for each single subband that is input to the prediction loop,
a synthesis filter bank is applied to the reconstructed subbands before the
motion compensated prediction is performed in the spatial domain. This
motion compensated prediction signal must then again be decomposed into
subbands before it is subtracted from the next incoming subband frame.
Thus, we observe that an additional filter bank, compared to what was

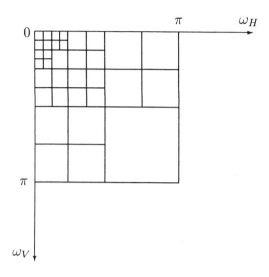

Figure 9.15: *Subband partitioning of [150].*

needed in the case when the subband splitting was done inside the prediction loop, is needed. This results in a coder of higher computational complexity.

The subbands are quantized using a uniform quantizer followed by an entropy coder. Only signal samples in subband blocks having undergone significant changes are sent to the decoder. To match this variable bit rate coder to a fixed rate channel, some kind of buffering is necessary. This is implemented by varying the "significant change" threshold of the quantizer in accordance with the instantaneous rate of bits produced by the coder and the buffer fullness.

Coding approaches similar to the present one were also presented by Biemond [127]. In this case the representation of the subband signals was accomplished by means of vector quantization. Kronander [93, 135] also presented a similar scheme, but the rationale for splitting the image frames into subbands was as follows: In the human visual system the sensitivity to different *temporal* frequencies varies depending on spatial frequencies in the image frames. Thus, by splitting the image frames into different spatial frequency channels (i.e. subbands), varying degrees of temporal filtering can be applied in the various spatial subbands. The possibility of selective temporal subsampling of various frequency channels following the temporal filtering was also considered in [93, 135].

Simplified Subband Coding with Motion Compensation

In [150] filter banks were introduced just before and just after the frame memory in the prediction loop in order to be able to apply motion compensated prediction in the image frame domain. This is, as Gharavi [134] observed, not the only way in which motion compensation can be applied for this type of video coder.

The obvious solution to the motion compensation problem when the subband splitting operation is applied directly to the image frames is to perform motion compensated predictive coding *separately* for each subband. This seems to be particularly appropriate in the context of image sequences with high spatial resolution. In [134] such a scheme is employed in the coding of 1920 × 1024 pel HDTV video conference sequences at about 5 Mbps. The image frames are split into 7 spatial subbands as shown in Figure 9.16. The characteristics of the lowpass band are similar to those of the original sequence at a lower resolution. The dimensions of this band are similar to those of the CCITT specified *common interchange format* (CIF), which is the format envisaged for input to low bit rate coders as specified in CCITT standard H.261 [86]. Therefore, a coder similar to the H.261 coding standard is proposed for this band. The other bands are coded using DPCM with motion compensation. Quantization and variable word length coding for these bands are done in a manner similar to what was described in subsection 9.3.1.

In applying motion compensation in this scheme, several alternatives were explored:

1. The motion estimates are obtained for each band and applied separately to each image sequence of subbands.

2. Motion estimates are obtained only for the lowpass band. These motion estimates are used also for the motion compensated prediction of the higher subbands.

3. Inter-frame coding with no motion compensation is used.

4. As a reference situation, straightforward *intra-frame* PCM coding of the subbands is employed.

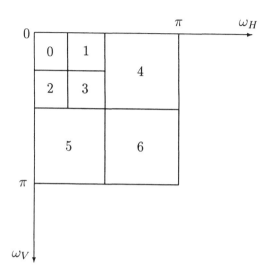

Figure 9.16: *Subband splitting employed in [134].*

It was found that for bands 3 and 6 (see Figure 9.16) intra-frame PCM coding, corresponding to case 4 above, was appropriate. For bands 4 and 5, inter-frame coding was found suitable, but motion compensation for these bands did not lead to any advantages. The remaining bands were found to benefit from motion compensation. However, bands 1 and 2 did not benefit from using the motion estimates obtained from these subbands. Instead it was found that the motion compensation was equally effective when using motion estimates obtained for the lowpass band. Thus, we see that the computational burden associated with motion estimation is reduced considerably at a negligible cost in terms of reduced performance. This coding scheme was claimed to produce excellent quality HDTV video conferencing images with 15 frames per second at 5.6 Mbps.

9.3.3 Improved Hybrid Motion Compensated Schemes

Many of the hybrid subband coding schemes incorporating motion compensated prediction are more or less straight forward adaptions of coders based on block oriented transforms, most notably the DCT. In [151, 152] several issues in subband video coding are addressed from the point of view of the inherent differences between subband decompositions and block oriented transforms.

The coder structure in [152] is conventional in the sense that it conforms to the traditional hybrid coding approach as exemplified by the model 1 coder introduced previously. The blocking effects of coders based on block transforms are sought removed by employing a *lapped orthogonal transform* (LOT), which is a particular kind of filter bank [75]. Young and Kingsbury [152] point out that also such coders may exhibit blocking artifacts *unless the basis vectors constituting the transform decays smoothly to zero at the endpoints*. This problem has been dealt with in Chapter 8 for still image coders. To avoid the blocking problem *modified* LOT's are designed such that the basis functions have no discontinuities at their endpoints.

A problem of video-coders conforming to the model 1 structure is the fact that in areas where traditional block matching with integer pel accuracy is inefficient, the prediction error is large but uncorrelated. These areas are difficult to code efficiently using any approach based on high sample to sample correlation. For this reason a loop filter is introduced to smooth out the predicted signal so as to make the prediction error more amenable to bit efficient representation. Unfortunately this results in undesirable blurring of the decoded image. As an alternative, attempts are made at improving the motion compensation by using subpel accuracy displacement estimates. In [152] this is done with a frequency domain motion estimation technique utilizing *overlapping* blocks. In doing this, windowing is applied to ensure that local pixels are given more weight than those further away. This gives more accurate motion estimates, resulting in a smaller prediction error, and consequently, lower bit rate.

The scheme described above is a pure inter-frame coding technique. In many practical coding algorithms conforming to the model 1 structure based on block transforms, each block is analyzed with the objective of determining whether or not this particular block is suitable for inter- or intra-coding. In the latter case the prediction loop is temporarily disabled. Jozawa and Watanabe [151] consider such a coding scheme. It is pointed out that this is a very natural thing to do in block based coders when the block transform and the blocks to be inter-/intra-coded are of equal size. In subband schemes this may pose problems because the filter responses are so long that adjacent blocks are not processed independently. Since there may be large discontinuities on the border between blocks that are destined for inter-

and intra-coding, artificial high frequency components are introduced into the prediction error signal when decomposed into subbands. This makes it more difficult to achieve the goal of bit efficient representation of the subbands. In fact it is shown that when no measure is taken with respect to these discontinuities, referred to as *level gaps*, a traditional DCT is actually preferable to some subband decompositions.

Because of the detrimental effect of these level gaps a model 2 type coder (see Figure 9.2 b) is put forward as an alternative. This model is, however, also abandoned as a result of the lower accuracy of the motion estimates for the motion compensation now taking place in the subband domain. This lower accuracy and the attendant larger prediction error are a direct consequence of the subsampling of the subbands. Instead the model depicted in Figure 9.17, where the original frames are decomposed into subbands while the motion estimation and prediction take place in the spatial domain, is put forward. Using this structure an LOT based signal decomposition was found to be slightly superior to a similar structure employing the DCT. It is interesting to note that the proposed structure of [151] as shown in Figure 9.17 is almost identical to the one introduced by von Brandt [150], see Figure 9.14. Are we back at where it all started?

9.3.4 Three-Dimensional Approaches

A digital image sequence is a 3-D signal with correlation in both the spatial and temporal directions. Following the logic of still image coding in which 2-D transforms and subband decompositions are applied, the sensible extension for image sequences would be to apply 3-D transforms or subband splitting operations, the outputs of which are represented in a bit efficient manner.

Given the fact that a DCT coder is considered as a particular case of subband coding it is interesting to review the results presented in a 1977 paper by Roese [153]. That paper reports coding results of a 3-D DCT coder for digital image sequences in which zonal sampling[7] is used. In this early study the 3-D approach was compared to a model 1 type hybrid scheme with a DCT in the spatial domain combined with motion compensated prediction along the time axis. It was concluded that the hybrid scheme was somewhat

---

[7]This corresponds to retaining quantized versions of transform coefficients in a particular region (zone) and discarding the rest [10].

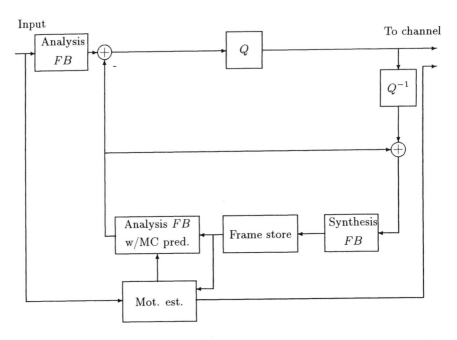

Figure 9.17: *Structure of the coder suggested by Jozawa and Watanabe.*

better. Such comparisons are, unfortunately, not given in any of the more recently introduced 3-D subband schemes for digital video. In the following we will cover two representative, and similar, schemes. First, a technique operating at 1 – 3 Mbps, intended for use in conjunction with integrating video into packet switched transmission networks, is presented. Following this, a low bit rate scheme (128 – 384 kbps) incorporating *structure vector quantization* will be described.

Medium Rate 3-D Subband Coding for Packet Transmission

In packet networks incorporating video services, data compression is highly desirable. It is recognized that hybrid coders with motion compensated prediction are not very robust with respect to loss of information. This poses particular problems in conjunction with video delivery over packet networks because of the finite probability of *packet loss*. To circumvent this problem a 3-D subband coder is proposed by Karlsson [138, 139].

In the present scheme the image sequence is split into subbands using a tree structured filter bank as shown in Figure 9.18. First the sequence is split into two temporal bands using the simplest possible nontrivial perfect reconstruction filter bank, namely the one using the filters[8]

$$H_{LP,t}(z) = \frac{1}{2}(1 + z^{-1}) \qquad (9.5)$$

and

$$H_{HP,t}(z) = \frac{1}{2}(1 - z^{-1}). \qquad (9.6)$$

The lowpass temporal band is subsequently split into 7 spatial bands,

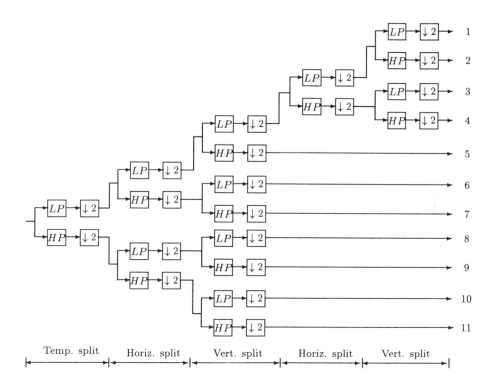

Figure 9.18: *Structure of 3-D spatiotemporal filter bank producing 11 subbands.*

whereas the temporal highpass band is split into 4 equal size bands. This

---

[8]Note that this selection of filters corresponds to a two point *Hadamard transform* in the temporal domain.

spatiotemporal subband splitting is depicted in Figure 9.19. The filters used
in the spatial domain filter bank are the low complexity FIR filters of LeGall
[142]. These analysis filters are given by

$$H_{LP}(z) = \frac{1}{4}(-1 + 2z^{-1} + 6z^{-2} + 2z^{-3} - z^{-4}) \tag{9.7}$$

$$H_{HP}(z) = \frac{1}{4}(1 - 2z^{-1} + z^{-2}). \tag{9.8}$$

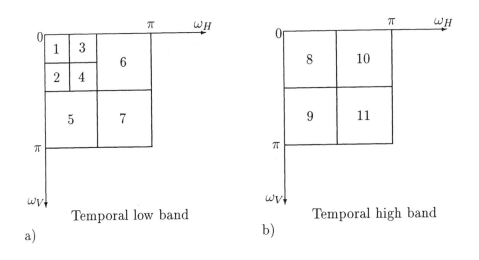

a)

b)

Figure 9.19: *11 band spatiotemporal subband splitting.*

In direct analogy with still image subband coding it is observed that the
spatiotemporal lowpass band is of primary importance. Also, the correlation properties of this band are reminiscent of those of the original image
sequence. This indicates that the lowpass band should be put into packets
with higher priority, to the extent that this is possible, than those used for
the remaining subbands. Furthermore, because of the remaining substantial
correlations in the lowpass band, a DPCM scheme is proposed for the coding of this band. To ensure a coder with low delay, the prediction is based
only on pels in the current lowpass frame. For error recovery purposes and
complexity reasons, a 1-D predictor is selected. The prediction error signal
is quantized with a uniform quantizer with an unbounded number of levels.

Regarding the perceptually less important higher subbands, a set of
quantizers optimized for the probability density function of the subbands

were first tried out. It was found, however, that the perceptually optimized quantizers of Gharavi [84], originally conceived in a still image coding context, gave visually more pleasing results.

Given the intended application of the present coding approach, the effect of packet loss was investigated. In the packetization stage, data for each subband are assembled into separate packets implying that the loss of a packet will only affect a single subband. For realistic packet loss probabilities it was found that lost packets corresponding to the higher 10 subbands have negligible perceptual consequences. The loss of packets corresponding to the spatiotemporal lowpass band was deemed to result in unacceptable perceptual degradations. Thus, investigation of schemes to minimize this effect was initiated. As a partial remedy, the idea of assigning higher priority to the packets corresponding to the lowpass band was put forward. Furthermore, in the event that these high priority packets are still lost, the possibility of replacing the lost subband data with subband data from previous frames was investigated. For areas with low to moderate motion this was found to be an acceptable solution. In areas of violent motion a better solution was to use a motion compensation scheme in selecting the previous subband data to use for the replacement.

Low Rate 3-D Subband Coding Employing Vector Quantization

In [136, 137] Podilchuk et al. studied a scheme similar to the one above for low rate coding in the 128 – 384 kbps range. The splitting into subbands is exactly as that of the previous section, i.e. 11 spatiotemporal bands are used. With respect to the filter bank, the only difference is the filters employed[9].

As in Karlsson's study the importance of the lowpass band is recognized. This prompted the application of a lossless compression scheme for this band. Regarding the higher subbands, they were judged to be *structured* and *sparse*. Therefore, a relatively crude approach to the representation of these subbands was suggested. First, for the bit rates considered, the 3 higher spatial bands of both the low and high temporal bands were completely discarded. This corresponds to setting bands no. 5, 6, 7 and 9, 10 and 11 of Figure 9.19 equal to zero[10]. The remaining subbands are coded using a

---

[9]The exact selection of filters is not stated explicitly in [136, 137].

[10]Also note that this is essentially the same as first performing spatial filtering and downsampling by a factor two on the original sequence to get a sequence of half the spatial resolution and then splitting this sequence into 5 subbands. That is, two temporal

computationally efficient vector quantization scheme that is optimized for preserving the underlying edge geometries of these subbands.

The vector quantization scheme for the higher frequency subbands that are not discarded works as follows: Rather than forming a codebook through a clustering algorithm, as is done for example by the LBG algorithm [60], a binary codebook is postulated at the outset. This is selected as 3×3 binary patterns representing observed structures in the higher subbands. An example of such a codebook with only 9 code words is shown in Figure 9.20. On the other extreme, a codebook consisting of all possible 3×3 binary patterns could also be used.

The patterns in the various subbands cannot be approximated in any reasonable way by these binary codebook vectors. Therefore, in searching the best codebook vector, reasonable grey level intensity values, I_w and I_b are assigned to the white and black elements, respectively, of the 3×3 binary patterns. Thus, in comparing a 3×3 block from a subband with a 3×3 binary codebook vector, one adjusts I_w and I_b so that the difference between the subband block and the intensity modulated codebook vector is minimal. This procedure is followed for all candidate codebook vectors, and the one giving the best match to the subband block is selected as the block's representative. The corresponding codebook index *along* with the best I_w and I_b values are sent to the decoder. The efficiency of the procedure, in terms of compression ratio, depends on the frequent selection of a small number of possible codebook vectors. This makes efficient entropy coding possible. In practice the null vector of constant intensity is the most frequently selected codebook vector.

This technique for image sequence compression was demonstrated successfully for simple head and shoulder images, typical of video-telephony. For simple sequences good results were reported at a bit rate of 128 kbps whereas for higher activity video-phone scenes 384 kbps was found to be a more appropriate bit rate.

9.4 Medium-to-high Rate Coding

Low to medium bit rate video coding primarily finds applications in visual telecommunications involving a limited number of communicators. These include video telephony, video conferencing, telemedicine as well as low-end

---

bands of which the lowpass band is split spatially into 4 bands.

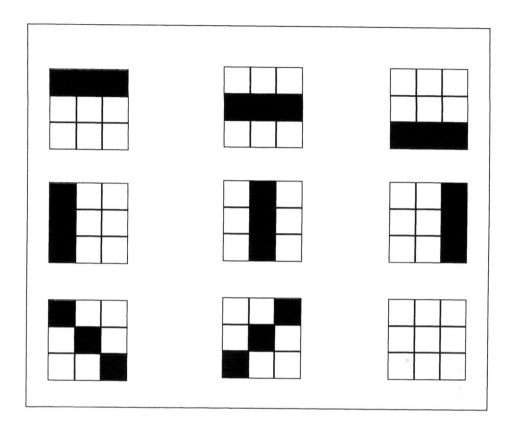

Figure 9.20: *Example of binary codebook of* 3 × 3 *patterns with* 9 *codewords.*

video on demand systems. Since these applications involve a small number of communicators, the cost, in terms of bit rates, must be kept at a minimum. In situations where the number of image consumers is larger the issue of quality plays a more prominent role than the issue of bit rate economy.

9.4.1 Background on Applications

In the near future several new services as well as enhancements of existing services involving moving imagery are envisaged. Examples are digital TV and high definition TV (HDTV) distributed via terrestrial radio links, satellite links or fiber based broad band ISDN (B-ISDN) facilities. Other applications include video-on-demand services via cable and packet-switched networks, storage of digital video on packaged media as well as high-end interactive services. Common to all these applications is the emphasis on high image quality. Still, the raw bit rates resulting from straightforward digitization of the source signals are so high, often more than one Gbps, that some sort of compression is necessary or highly desirable.

In conjunction with the terrestrial distribution of digital high definition and enhanced definition TV signals (HDTV or EDTV), the challenge is to compress the digital video with very minor degradations in such a way that the bit-stream can be modulated to fit into the existing 5 – 6 MHz terrestrial and cable channels. In terms of bit rates this translates into 15 – 30 Mbps when established modulation techniques are used. Employing sophisticated modulation techniques, even higher bit rates might be possible for terrestrial EDTV/HDTV broadcasting. It should be mentioned in this context that pure analog approaches to the representation of EDTV/HDTV signals have been favored by some until quite recently. For a brief description of such approaches see [154].

In the context of digital HDTV distribution via B-ISDN - the envisaged all purpose exchange area communications vehicle in the not too distant future, coding at bit rates of about 140 Mbps and 35/45 Mbps, has received considerable attention. The main reasons for the focus on these bit rates are the requirements for superior picture quality and the provision of these bit rates in the digital transmission hierarchy. Another important consideration related to the distribution of compressed digital HDTV signals is the issue of compatibility. The compatibility issue can be stated as follows: It is reasonable to assume that TV receivers of different capabilities with respect to spatial image resolution will be coexisting. It is therefore important that within the HDTV bit-stream complete bit-streams for one or more lower resolution images are embedded in such a way that also cheaper receivers not supporting full resolution HDTV can be used. In the following subsections the following main topics will be covered:

- Intra-frame HDTV coding at 140 Mbps.

- Hierarchical subband techniques at 140 Mbps explicitly addressing the compatibility issue.

- Subband video coders operating at 35/45 Mbps.

Name	Res. ($l \times c$)	Bit rate (Mbps)
HDP (High definition progressive)	1250×1920	1728
HDI (High definition interlaced)	1250×1920	1152
EDP (Enhanced definition progressive)	625×960	432
EDI (Enhanced definition interlaced)	625×960	288
(HQ)VT ((High Quality) Video telephony)	312×480	108

Table 9.1: *A possible hierarchy of digital TV formats adapted from [155]. Note that the raw bit rates include both luminance and chrominance information, the latter subsampled by a factor of two relative to the former vertically.*

- Lower rate techniques for terrestrial broadcast and other applications (MPEG 2).

9.4.2 Intra-Frame Subband Coding for HDTV

Initial focus on HDTV image compression were on intra-frame approaches with target bit rates around 140 Mbps. Reasons for the emphasis on intra-frame as opposed to inter-frame coding was to a large extent related to complexity considerations. Also, in studio applications where the independent decoding of arbitrary single frames is important, intra-frame coding is attractive. Although many researchers are moving towards more computationally demanding inter-frame schemes, mention of the intra-frame approaches is warranted.

The target bit rate of 140 Mbps should be considered relative to the raw bit rate of a digital HDTV signal. Unfortunately there is, at present, no agreed upon world wide digital HDTV format. One proposed hierarchy of digital TV formats including HDTV is reproduced in Table 9.1. From this table we see that taking the HDP format as a reference, a compression ratio of 16:1 is required in order to fit a raw digital HDTV signal onto a 140 Mbps transmission channel.

In [156] an intra-frame subband coding scheme for HDP signals is proposed. This is a "pure" subband coding scheme in the sense that the subband signals are represented through a simple PCM encoding strategy. Initially a 7 band split as shown in Figure 9.21 a) is investigated. The rationale for this splitting is the relation between subbands $\{1\}$, $\{1,2,3,4\}$ and $\{1,2,3,4,5,6,7\}$ to the VT, EDP, and HDP signals, respectively. Applying independent quantization to the subbands, a compatible coding system results. Sufficient quality is, however, not obtained using this split: For this situation it was found that the 16 band split of Figure 9.21 b) was preferable. Still, it is argued that the quality of the decoded HDP signals is not sufficient for a targeted bit rate of 140 Mbps. Based on this the authors end up recommending an inter-frame strategy. More will be said about this later.

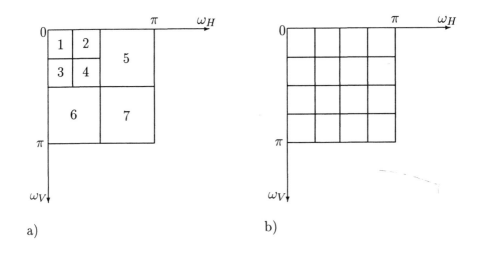

Figure 9.21: *a) 7 and b) 16 band splits in the compatible coding scheme of [156].*

Ansari et al. [157, 82, 154] also describe an inter-field system for HDTV signal compression at 140 Mbps for transmission over B-ISDN facilities. Employing raw image data roughly corresponding to the HDI format of Table 9.1, subband coders with computationally efficient IIR structures as well as short kernel symmetric FIR filter banks are proposed. In both cases a simple 4 band separable split is employed. The low pass subband is further coded using a DCT based coder similar to the scene adaptive coder of Chen and Pratt [85]. For the 3 higher frequency bands the sample to sample correlation was deemed low enough to justify a pure PCM coding scheme. Several quantizer characteristics for this PCM coder were examined. It was found that a uniform quantizer with a large dead-zone (see Figure 4.2) performed best. The dead-zone is introduced to ensure a high frequency of zero-levels in the higher subbands. This high frequency of zeros makes these subbands amenable to subsequent run-length coding.

The filter banks used in the work described above were based on very simple IIR or FIR filters. The selection of simple filters was justified by the high data rates of digital TV signals and the implementability of circuitry at such rates. In [158] some evidence is presented to support the claim that, at the bit rates considered here (1.5 bit per pel or more for the luminance component), the exact characteristics of the filter bank employed in a coder is of modest importance. In fact, the performance of simple low complexity filter bank structures are on par with more computationally demanding ones.

The HDTV compression schemes described above produce variable bit rates. The bit rate produced at any given time depends on the complexity of the image frame to be coded at that particular moment. Although a variable bit rate service may be foreseen in a B-ISDN context, constant or constrained bit rate services are likely to prevail. In conjunction with the present coding approach, ways of controlling the bit rate were proposed. First, in order to ensure that the bit rate of the video coder does not exceed the allotted channel capacity, a buffer scheme must be implemented. Depending on the fullness of this buffer, several parameters of the coding algorithm may be adjusted to match the bit production rate to the target bit rate. In [154] two approaches to the channel rate adaption problem is presented. First, a variable prefilter is proposed. The purpose of this prefilter is to smooth the image field to be coded so as to make it less demanding for the coding algorithm. As the buffer fills up the degree of lowpass filtering is increased. If this mechanism of rate control is employed alone there is a danger that particularly demanding image sequences may be subjected to unacceptable softening. Therefore, adaption of the quantizer steps of the high band PCM coders should be employed in addition to the variable prefiltering. This adaption implies that as the buffer fills up the quantizer step size is increased so that the instantaneous bit rate decreases. A suitable combination of these two rate control mechanisms is claimed to be superior to either one alone.

Regarding the quality of the decoded video at bit rates corresponding to about 140 Mbps, it is judged as "pleasing" both spatially and temporally. Differences between original images and decoded images were, however, observed on high quality monitors. The distortion appeared mainly as a slight ringing noise close to high contrast edges. The exact nature of this distortion will depend on the exact choice of filter bank.

Intra-frame HDTV coding below 140 Mbps At lower rates than about 140 Mbps, intra-frame coding as described above results in unacceptable image quality. Subsampling using a quincunx downsampler (see Figure 9.22) is proposed as a good way of producing a lower resolution signal which is subsequently coded using a subband coder with nonrectangular frequency partitioning [154]. The rationale for using a quincunx downsampler is that the information in most natural imagery is mainly confined within a diamond shaped region of the spatial frequency domain. This corresponds to the cross-hatched portion of Figure 9.22 a). A suitable resampling lattice in this case is the quincunx lattice [2], which is illustrated in Figure 9.22 b). Prior to the downsampling a diamond shaped filter is used to discard frequency contents outside the cross-hatched area of Figure 9.22 a). This area is subsequently split into nonrectangular subbands as depicted in Figure 9.23 a). Note that the frequency band labeled HH has already been discarded as a result of the prefiltering/quincunx resampling. The LL and LH bands, available on the rectangular sampling grids, are further processed by DCT and PCM coding systems, respectively. The coding results are claimed to be good[11], but some lowpass effects are apparent due to the discarding of some high frequency information.

9.4.3   Compatible Subband Coding of HDTV

Subband coding is inherently a *hierarchical* coding method in the sense that a decomposed image provides the possibility of reconstructing images at different levels of spatial resolution. This is accomplished by selecting various subsets of the subbands for the reconstruction process. In commercial utilization of digital HDTV signals it is considered mandatory that the signals be useful also for receivers providing lower spatial resolution than full HDTV. Thus, in addition to providing high quality service at full HDTV resolution

---

[11]We have yet to come across some coding publication not claiming this!!

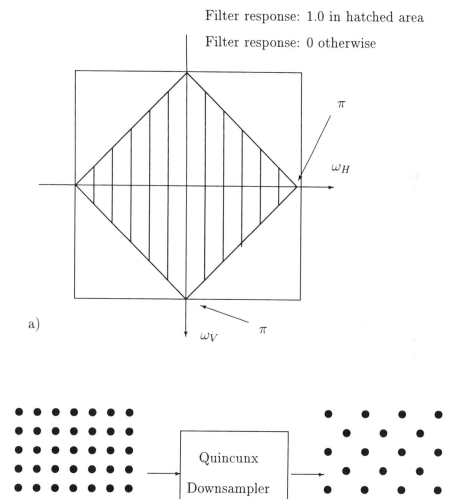

Figure 9.22: *a) Frequency response of ideal diamond filter, b) Quincunx resampling.*

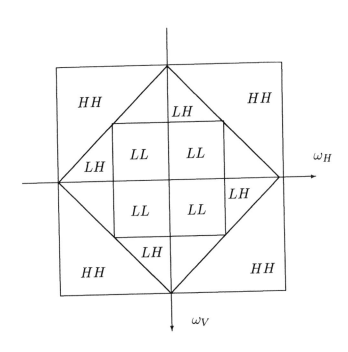

Figure 9.23: *Frequency band partitioning resulting from diamond shaped pre-filtering, quincunx resampling, and nonrectangular two-band decomposition.*

at about 140 Mbps, embedded bit-streams decodable for lower resolution receivers must be provided for. Ways of catering to this need is the topic of the present subsection. First we deal with adaptions of pure intra-frame subband schemes. Subsequently, recently introduced hybrid schemes involving both subband decomposition and motion compensated prediction are presented.

Intra-frame compatible subband coding of HDTV   The material in this subsection is based mainly on the work of Bosveld, Biemond and colleagues [159, 160, 161, 162]. Related and complementary issues to what is presented here can be found in these references.

In a hierarchical coding system for HDTV signals various possibilities regarding the number of embedded lower resolution signals exist. On the extreme side one could consider a full hierarchy having embedded bit streams for all the levels of Table 9.1. Such a hierarchy is considered in [162]. In that work a full spatiotemporal hierarchy based on a 3-D nonseparable subband decomposition was considered. This full hierarchy was compared to a so-called reduced temporal hierarchy in which all interlaced formats of the aforementioned table are converted to a progressive format prior to coding, and, if appropriate, back to interlaced format after decoding. Somewhat surprisingly it was found that the reduced temporal hierarchy was superior to the full hierarchy including interlaced formats.

Focusing on the important issue of allocating bits to the various resolution signals in a compatible coding system we concentrate on a system involving two distinct resolution levels. In doing so we treat subband coding of HDTV at 135 Mbps with an embedded 35 Mbps signal providing EQTV resolution (HDP and EDP in Table 9.1). Bosveld et al. [160] consider three distinct approaches to this problem: *distributed coding, error feedback coding* and *selective coding*.

In all three cases a tree-structured filter bank using the 32 tap QMF filter designated as 32C by Johnston [17] is employed. The image frames are split into 28 subbands. This is the same frequency band partitioning as suggested by Westerink [77] for a pure still image coder. The decomposition is depicted in Figure 9.24. Note that the 16 subbands in the upper left corner of Figure 9.24 correspond to the EQTV signal, whereas the remaining subbands add the HDTV resolution. For this reason the signal made up from these latter subbands is called the HDTV complement or HDTVc for short.

In a system employing *distributed coding* the EQTV subbands and the HDTVc subbands are coded independently. That is, for the EQTV signal the bit rate of 35 Mbps is assigned and a bit allocation algorithm distributes the bits in an optimal way to the 16 subbands. These are subsequently quantized with quantizers optimized with respect to the *probability density function* of the subbands (pdf-optimized quantizers). The quantizer outputs are furthermore represented by a Huffman code. To provide a total bit rate of 135 Mbps in the complete HDTV signal, the remaining 100 Mbps is spent on the HDTVc subbands. These subbands are quantized and coded according to the same procedure as described for the EQTV subbands. We illustrate the principle of distributed coding in Figure 9.25.

It is well known that successful subband coding is critically dependent on the accurate representation of the lower subbands. The problem with

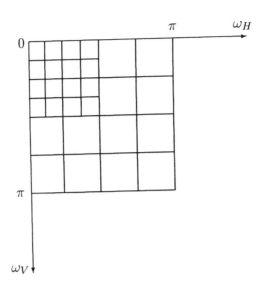

Figure 9.24: *28 band split used in [160]. for a two resolution compatible HDTV coding system.*

distributed coding in a 35/135 Mbps context is related to the following: Although the EQTV component is coded optimally with the given bit rate constraint, the accuracy in the representation of these subbands is not sufficient for the full HDTV image. Given the 135 Mbps constraint for the full HDTV signal it would have been preferable to spend more than 35 Mbps on the 16 subbands constituting the EQTVs signal.

The *refinement system* represents one possibility for dealing with the deficiencies of the distributed coding system. In this system the EQTV signal is first coded as described above. Thus, this component of the signal is still optimally encoded. The distinguishing feature of the present strategy is that, rather than spending all the remaining bits on the HDTVc subbands, the difference signal, formed by subtracting the uncoded EQTV subbands from the decoded EQTV subbands, is allowed to compete for the remaining 100 Mbps along with the HDTVc subbands. This ensures that the EQTV subbands as used in the reconstruction of the full HDTV resolution signal are represented accurately enough for adequate quality of the full resolution signal. Thus, in this approach the encoding of the HDTVc subbands must

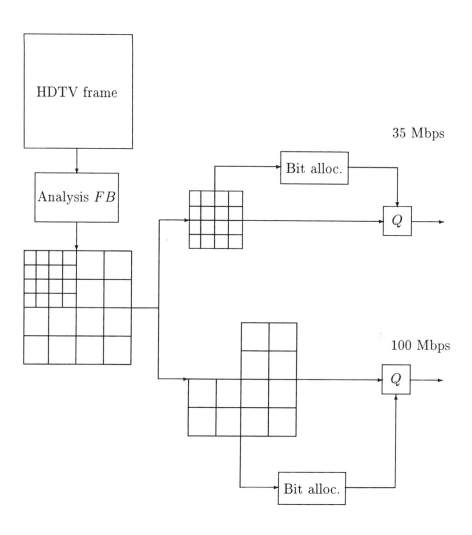

Figure 9.25: *Distributed coding of HDTV subbands at 35 Mbps (EQTV) and 135 Mbps (full HDTV).*

be performed after the encoding (and local decoding) of the EQTV subbands has been completed. A block diagram illustrating the principle of refinement coding is shown in Figure 9.26.

Although the EQTV portion of the signal is optimally encoded in the refinement system - there is no guarantee that this is also the case for the complete HDTV signal. In the so-called *selection system* - see Figure 9.27, all the 28 subbands are allowed to compete for the total rate of 135 Mbps through the bit allocation stage. This ensures optimal encoding of the full resolution HDTV signal given the constraint on the total bit rate. For the 16 EQTV subbands this procedure typically leads to a total bit assignment exceeding the 35 Mbps constraint. This problem is solved by selecting a subset of the 16 EQTV subbands to be used for the lower resolution signal. Some of the high EQTV resolution subbands that do not fit into the 35 Mbps bit stream are sent along with the HDTVc subbands.

In [159, 161] a thorough theoretical analysis of these schemes, founded on rate distortion theory and some idealized assumptions, is presented. Based on the theoretical analysis and experimental evidence it is concluded that the refinement system is superior to the two other alternatives considered for the given distribution of bit rates in the composite bit stream. In particular, when comparing the refinement and the selection systems for the low resolution signal, it was found that refinement system was preferable. With respect to the full resolution HDTV signal the selection system was preferable, but only marginally so. From this it was concluded that the refinement system was the preferred one of the three alternatives examined.

Inter-frame compatible HDTV coding   The compatibility problem for intra-frame HDTV coders essentially amounts to the appropriate distribution of bits among the various bit streams corresponding to the various image resolution classes. Except for this, these coders are essentially high rate still image coders. Although intra-frame techniques are attractive from the point of view of low hardware complexity, it is evident that the unexploited correlation along the time axis results in suboptimal coding performance. Given this, the adoption of some hybrid scheme, as described earlier in this chapter, including motion compensation, has received considerable attention also in an HDTV context. The superior performance of such schemes is paid for by more complex/expensive encoding/decoding hardware relative to what is required in pure intra-frame approaches. Also, the use of motion

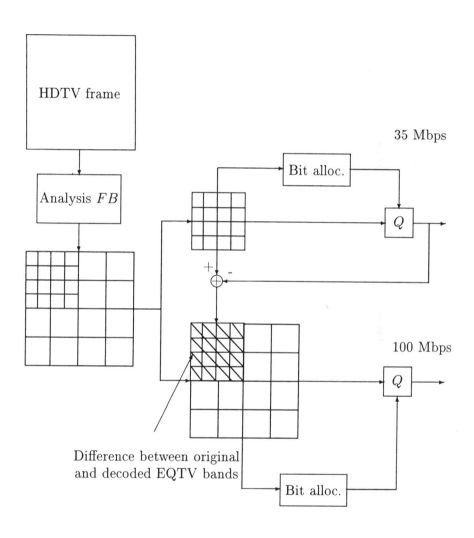

Figure 9.26: *The refinement coding system for two level compatible HDTV coding.*

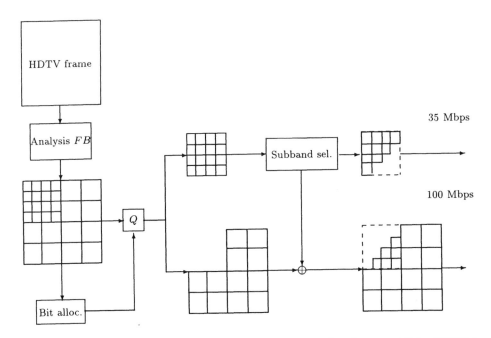

Figure 9.27: *The selection coding system for two level compatible HDTV coding.*

compensated coders in a compatible context poses some specific problems in need of a solution. Some of these issues will be presented in the following paragraphs.

As has been stated before, hybrid coders involving subband decomposition perform this decomposition either inside or outside the prediction loop (i.e. either each prediction error image or each original image frame are split into subbands). The most commonly used approach in low rate applications is inside loop subband decomposition. This is so even though certain objections have been put forward: The prediction error image in a coder with a block oriented inter/intra-mode switch and possibly block oriented motion compensation may have substantial discontinuities at the block boundaries that are not handled well by filter banks whose constituent filters have unit pulse response lengths exceeding the size of these blocks. Exactly the same argument has surfaced in conjunction with hybrid coding schemes for HDTV [163]. For this reason and for reasons of compatibility the major focus has been on schemes in which the image frames are split into subbands prior to the prediction loop.

The conceptually simplest scheme referred to by both [155] and [163] as the *in-band motion compensation scheme* is depicted in Figure 9.28. In this scheme each frame is decomposed into a number of subbands, for example 64. Each subband signal is subsequently coded by a motion compensated predictive coder. So far this structure looks very similar to the model 2 structure of Gharavi [115] described in Section 9.3. In that model, however, the motion estimation was performed on a subband basis only. In the compatible HDTV coders of [155, 163], the motion vectors are estimated from two subsequent original *full resolution* frames. The motion estimates are therefore more accurate than they would have been when estimated from the subbands. Given an integer pel accuracy for the motion vectors in the original full resolution image domain, we get subpixel accuracy in the decimated subband domain. There is therefore a need for interpolation in the subband domain when the motion compensated predictions are to be computed. Unfortunately, this is complicated by the fact that aliasing is introduced by the subsampling operation involved in the formation of the subbands. What is really needed in performing the interpolation is access to the subbands prior to the subsampling. This in turn requires knowledge of all the other subsampled subbands. Performing this interpolation is of course a demanding computational task. In [163] it is shown how this interpolation can be performed approximately at a reasonable computational cost when it is observed that only the neighboring subbands need be included in the computations. To avoid violating the compatibility requirement, it is necessary that for a given subband, only subbands belonging to the same compatibility class as the given subband be included in this interpolation[12]. In [155] simple interpolation based only on information within each subband is utilized.

The other motion compensated hybrid technique we shall consider here is termed the *cumulative prediction method* in [155]. The technique is also described in [164, 163]. The main feature distinguishing this scheme from the *in-band* scheme above is the domain in which the motion compensated prediction is performed. In the present scheme the prediction is performed in the spatial rather than in the subband domain. An example of this for a

---

[12]For example: For a 64 band split where the 16 lower bands correspond to the EDP resolution level, we cannot make use of the higher bands belonging to the HDTV complement in performing the interpolation of the EDP bands because these bands are not available in a pure EDP decoder.

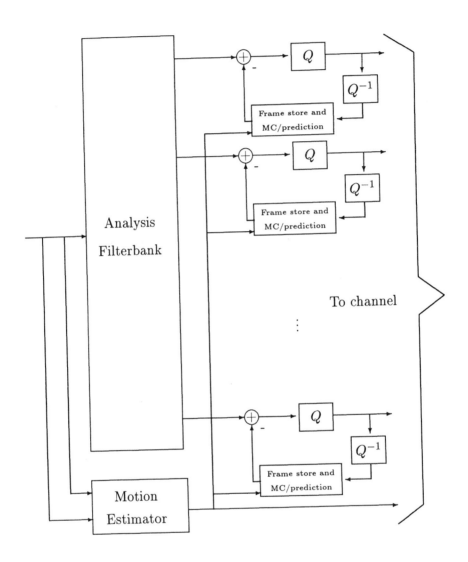

Figure 9.28: *Encoder portion of an* in band coding system. *Each subband is coded with a motion compensated coder utilizing motion estimates obtained from full resolution input frames. Since this implies subpel accuracy for the motion vectors in the subband domain, interpolation is necessary in performing the motion compensated prediction.*

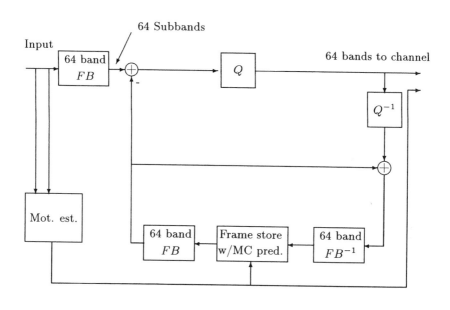

Figure 9.29: *HDP encoder with motion compensation.*

64 band subband coder is shown in Figures 9.29 and 9.30. Assume by way of example that we are dealing with a compatibility requirement involving the HDP and EDP signal formats.

In Figure 9.30 a) and Figure 9.30 b) two possible decoders operating at full resolution and at EDP resolution are also indicated. The HDP decoder in Figure 9.30 a) works fine *since the same information is used in both the encoder and decoder predictors.* The EDP decoder will, however, not work well since there is a mismatch in the information being utilized by the encoder and decoder predictors. The encoder uses all the subbands, whereas the decoder is only able to utilize information embedded in the lower 16 subbands. This mismatch between encoder and decoder affects the decoded picture quality significantly.

Assuming a system with 64 uniform subbands and an HDP/EDP compatibility requirement, the EDP receiver will base its reconstructed signal on the 16 lower frequency subbands, whereas the HDP receiver makes use of all the 64 subbands. To ensure that the predicted subbands are the same in the encoder and the decoder at both resolutions, we must make provisions in the

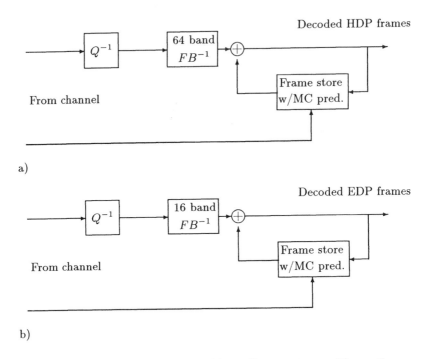

Figure 9.30: *Example of a noncompatible coding system with motion compensation. a) HDP decoder. b) Possible EDP decoder utilizing the 16 lower subbands in the reconstruction. This decoder does not work in conjunction with the encoder in Figure 9.29 since the prediction loop in the encoder makes use of information unavailable to the predictor in the decoder. Thus, there is a mismatch between the predicted lower 16 subbands in the encoder and the decoder. The mismatch between the predicted signals in the encoder and decoder makes this system unusable in a compatible coding context.*

encoder for motion compensating two prediction signals, one based on EDP bands and one on all the subbands. The structures of the encoder and the decoder in this case are shown in Figure 9.31 and Figure 9.32. Again, the integer accuracy motion estimates are based on two original frames giving rise to 1/2 pel accuracy in the motion estimates for the EDP loop. This again necessitates interpolation. Also note that the reconstructed subbands of the EDP loop are fed into the synthesis filter bank in the HDP complement loop so as to make prediction in the HDP loop dependent on *all* the subbands.

In this scheme the problem of how to spend the bit rate dictated by the channel bandwidth, as treated in detail for intra-frame HDTV coding schemes, resurfaces. In both [155, 163] an adaption of the refinement system of Bosveld [160] is successfully applied.

Vandendorpe et al. [155] give some simulation results for the purpose of comparing intra-frame coding, the hybrid motion compensated coding using the in-band technique and the cumulative approach just described. For the three image sequences tested it is evident that hybrid motion compensated schemes are far superior to an intra-frame scheme. It is furthermore shown that for sequences with little motion, the in-band and the cumulative schemes have comparable performances. In the case of image sequences with significant motion, the cumulative scheme is better. The failure of the in-band system in this case, is attributed to the poor interpolation as required by the subpel resolution of the motion vectors in the subbands.

In conclusion, the cumulative system is the preferred one among the schemes mentioned. For this scheme it was also found that the performance loss (for the HDP signal) relative to a noncompatible coder for only the HDP resolution was small.

9.4.4 35/45 Mbps Coding of Digital TV

In the future omnipresent fiber based B-ISDN environment, 45/35 Mbps and 140/135 Mbps are probable channel capacities available to the public. As evidenced in the previous subsection much effort has been put into 140/135 Mbps coding of HDTV. Some work on the coding of digital TV at 45/35 Mbps has also been reported recently. As a representative example we present the work of Hang and colleagues [165].

In [165] it is observed that the use of standard coding algorithms for HDTV images implies a demand for very high-speed hardware. This in turn leads to high cost coders. It is argued that parallel configurations of industry standard hardware might be a cost effective solution. In this context the standard hardware envisaged are the constituent parts of an H.261 coder. Rather than code the whole HDTV resolution picture as a single entity the basic idea is to partition the image into disjoint subimages that are independently coded with an H.261 coder. The partitioning is performed in two alternative ways:

1. An HDTV image is simply split into 12 subpictures directly in the spatial domain.

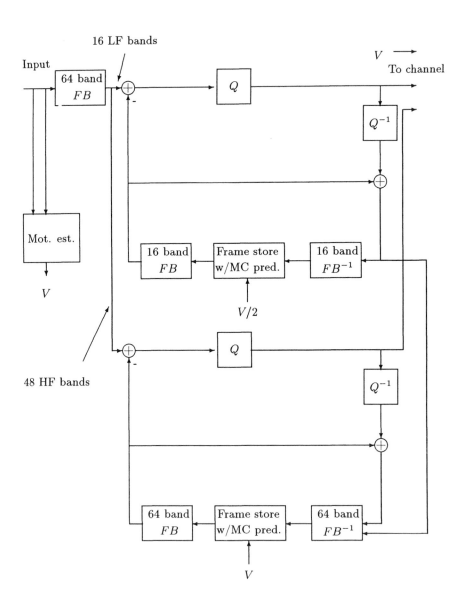

Figure 9.31: *Encoder of a* cumulative coding system *with an EDP/HDP compatibility requirement employing a 64 band subband decomposition.*

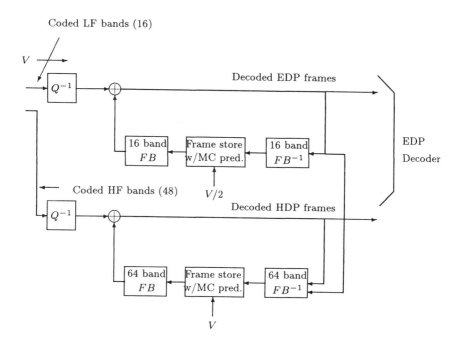

Figure 9.32: *Cumulative coding system; decoder. The full HDP decoder is shown with the EDP portion indicated.*

2. The HDTV image frame is split into 16 subimages, each corresponding to a different frequency range. This corresponds to a subband decomposition of the image.

In both cases the major problem is to devise a strategy for allocating bits to the various subchannels. In general terms this is done by attempting to distribute the bits in such a way that the distortion in each subimage is equal. This is done by establishing a model for the relation between the bit rate and the distortion for every subcoder. This model is used for dynamically computing a *channel sharing factor*, i.e. a number indicating the proportion of all the bits to be allocated to each subimage.

In the experiments reported in [165] the decomposition with subband filters was found to be inferior to the simple spatial splitting of the image. The reasons for this are:

• Lower accuracy motion estimates because the motion vectors are based on subbands, i.e. subsampled versions of the incoming image sequence.

- The H.261 coder used for each subchannel requires input in an 8 bit format. Thus, each subband was scaled and quantized to conform to this requirement. This process introduces additional errors.

- The algorithm for allocation of bits to the various channels was not deemed suitable for the large differentials in dynamic range among the various subbands.

It is pointed out that several of these problems should be rectifiable, at the expense of added computational complexity.

9.4.5 Standard TV Subband Coding at 4 − 9 Mbps

Any treatment on medium to high rate image sequence coding using subband techniques would be incomplete without at least some brief mention of the activities behind MPEG. In 1988 ISO/IEC formed the Motion Pictures Expert Group (MPEG) with the mandate of initially devising an international standard for the coding of video conforming to the CIF (common interchange format) at bit rates around 1.5 Mbps. One of the main applications considered at the time was storage on packaged media such as CD-ROM. The resulting recommendations are commonly referred to as MPEG-1. As this work progressed it was felt that a standard for higher quality images aimed at serving a broader range of applications was needed, thus the work on MPEG-2 was initiated. Several research laboratories all over the world have participated in this work aimed at devising algorithms for high quality compression of CCIR 601 image sequences at target bit rates of 4 and 9 Mbps. A large number of proposals were subjected to a wide range of subjective assessments in Kurihama in Japan in November 1991 [166]. Several of these algorithms are described in a 1993 special issue of *Image Communications* [167].

The majority of the proposals for MPEG-2 can be classified as different extensions of the MPEG-1 algorithm implying that they are based on the DCT. Proposals based on subband and wavelet decompositions were, however, presented. In this context we briefly mention the *subband based TV coding* scheme of Schamel and De Lameillieure [168]. In this coder the interlaced CCIR 601 fields are merged to form a 25 frames/sec. input signal. The signal is split into 16 uniform subbands using prototype FIR filters with few taps. Motion compensation is performed in the spatial domain in much the same way as in von Brandt's original subband coder as described previously. A scanning procedure is applied both within each subband and across the subbands for the purpose of creating 1-D scan strings that after quantization are amenable to run-length and entropy coding. It is concluded that this scheme exhibit a performance comparable to DCT schemes.

9.5 Summary

In this chapter we have presented a multitude of subband video-coders. Some of the differences between the various algorithms are targeted bit rate, computational complexity, number of subbands, type of filter banks, approach to the bit efficient representation of the subband signals, outside or inside loop signal decomposition, and the way in which motion compensation is applied. Again, within the confinements of each of these distinguishing features a multitude of choices can be made. Therefore, making clear-cut judgements about performance and suitability of a given coder for a given purpose is not easy. As is illustrated by the tests at Kurihama [166] it is not in any way obvious that the question "Should I use a subband coder or a DCT coder?" is particularly relevant. Too many factors beside this single issue enter into the question of "What is the best video coding scheme?". By way of example we mention the work on Subband HDTV coding [169] in which excellent coding results are obtained by using a 4 band uniform and separable spatial decomposition employing the filters $H_{LP}(z) = \frac{1}{2}(1 + z^{-1})$ and $H_{HP}(z) = \frac{1}{2}(1 - z^{-1})$. This illustrates in a clear-cut fashion that it is not necessarily the number of subbands nor the quality – in terms of transition width or stopband attenuation, of the filter banks that determine the overall suitability of the video subband coder. A complete and all-encompassing evaluation with the aim of finding the optimal video coder has yet to be performed.

Appendix A

Some Digital Signal Processing Facts

This appendix presents some useful background on allpass filters, anticausal filtering and multirate digital signal processing that is used in the main text.

A.1 Allpass Filters

An allpass filter has a magnitude response equal to unity for all frequencies:

$$|A(e^{j\omega})| = 1. \qquad (A.1)$$

Using polar form notation, the complete frequency response can be written

$$A(e^{j\omega}) = e^{j\phi(\omega)}, \qquad (A.2)$$

where $\phi(\omega)$ is the *phase* response of the filter. Since $\phi(\omega)$ usually is nonlinear, different frequencies will experience different amounts of delay implying that the form of the time domain signal changes.

The general transfer function of a digital allpass filter can be expressed as

$$A(z) = \pm \prod_{i=1}^{m} \left(\frac{z^{-1} + \lambda_i^*}{1 + \lambda_i z^{-1}} \right), \quad |\lambda_i| \leq 1. \qquad (A.3)$$

In this equation the *poles* are given by $-\lambda_i$. Writing the complex number representing the pole in polar form, $-\lambda_i = \rho_i e^{j\theta_i}$, there is always, for the

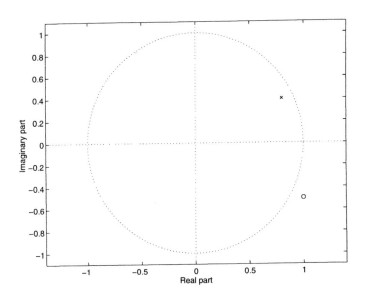

Figure A.1: *Pole-zero pair that constitute one factor in an allpass transfer function.*

case when $A(z)$ has real valued coefficients, a companion zero at the location $-1/\lambda_i = \frac{1}{\rho_i}e^{-j\theta_i}$. An example pole-zero pair is shown in Figure A.1.

The proof of the allpass property can be performed on each of the factors in the transfer function in Equation A.3. Writing the squared magnitude of the allpass frequency response of the ith factor as

$$|A_i(e^{j\omega})|^2 = A_i(e^{j\omega})A_i^*(e^{j\omega}) = \frac{e^{-j\omega} + \lambda_i^*}{1 + \lambda_i e^{-j\omega}}\frac{e^{j\omega} + \lambda_i}{1 + \lambda_i^* e^{j\omega}}, \qquad (A.4)$$

and multiplying the first factor of this equation by $e^{j\omega}$ in both numerator and denominator, the condition of Equation A.1 is established.

If the coefficients of the allpass filter are real, it is a direct consequence of Equation A.3 that

$$A(z)A(z^{-1}) = 1. \qquad (A.5)$$

An interesting property of the unit sample response, $a(l)$ of an allpass filter, is that it is orthogonal to itself for all nonzero shifts, or expressed mathematically

$$\sum_{l=-\infty}^{\infty} a(l)a(l+k) = \begin{cases} 1, & k = 0 \\ 0, & k \neq 0. \end{cases} \qquad (A.6)$$

This is not proven here. It can be used to find orthogonality properties of allpass based filter banks.

A.2 Anticausal Filtering

Some types of filter bank systems require the use of unstable filters. This is common for example in polyphase realizations of perfect reconstruction analysis-synthesis systems, where the polyphase filter in an analysis branch is stable and the polyphase filter of the corresponding branch of the synthesis filter bank is required to be the inverse filter of the analysis polyphase filter. This situation corresponds to the one depicted in Figure A.2a) where we have a cascade of a filter, $P(z)$ and its inverse $P^{-1}(z)$. Here we assume that $P(z)$ has all its zeros outside the unit circle, i.e. it is a maximum phase filter. Since $P^{-1}(z)$ is unstable, this cascade will not work, even for finite length signals. When filtering signals of finite length, this problem can be solved by reversing the signal to be passed through the unstable filter, filter this reversed signal through $P^{-1}(z^{-1})$ – which *is* stable – and finally reverse the output. This situation is depicted in Figure A.2 b). Thus, it possible to get an identity system while avoiding stability problems. The price for this is a delay equal to the length of the finite signal to be filtered.

A.3 Multirate Digital Signal Processing

Digital signal processing systems in which more than one sampling rate is involved are called multirate systems. Filter banks used in subband coders are examples of such multirate systems. In this appendix we present some essential facts of multirate digital signal processing theory.

A.3.1 Downsampling

An N-fold downsampler, or subsampler, is an operator that, based on an input signal $x(n)$, produces and output signal $y(m)$ consisting of every Nth sample of the input, i.e.

$$y(m) = x(mN). \tag{A.7}$$

In Figure A.3 we show a block diagram of an N-fold downsampler as well as an example of its operation for the case when $N = 2$. Subsampling of a signal $x(n)$ may be interpreted as a compression in the time domain. Thus,

a)

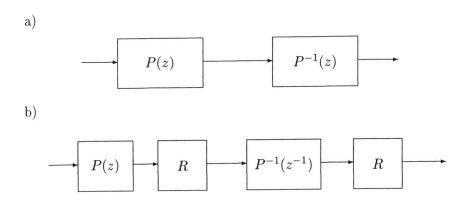

b)

Figure A.2: a) Identity system of a filter and its inverse in cascade. If $P(z)$ is not minimum phase, the system will not work, due to stability problems. Assuming that $P(z)$ is a maximum phase filter, the stability problem can be solved by anticausal filtering as depicted in b). The boxes labelled R, indicate that the finite length signals at these points are reversed.

one would expect an expansion in the frequency domain. In fact, it can be shown [11] that the z-transform domain counterpart of Equation A.7 is

$$Y(z) = \frac{1}{N} \sum_{k=0}^{N-1} X(z^{1/N} e^{-j\frac{2\pi}{N}k}).$$ (A.8)

The frequency domain expression is found by setting $z = e^{j\omega}$:

$$Y(e^{j\omega}) = \frac{1}{N} \sum_{k=0}^{N-1} X(e^{j(\omega-2\pi k)/N}).$$ (A.9)

For the case of $N = 2$ the above equation becomes

$$Y(e^{j\omega}) = \frac{1}{2}[X(e^{j\omega/2}) + X(-e^{j\omega/2})].$$ (A.10)

Example plots of the Fourier transform of the input and output signals are shown for two different cases in Figure A.4. Note that in the b)-part of the figure the two terms of Equation A.10 overlap. This is defined as aliasing, and precludes the possibility of recovering the original input signal. In the other case, however, there is no aliasing since the input is appropriately

a)

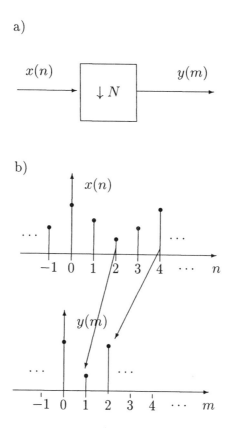

b)

Figure A.3: *Downsampler: a) Block diagram. b) Example of operation for N = 2.*

bandlimited prior to subsampling. Because aliasing is unacceptable in many applications, it is common practice to precede a downsampler by a digital lowpass filter for the purpose of bandlimiting the signal to be subsampled to frequencies below π/N. The combination of such a filter and a downsampler is called a *decimator*.

Another important example of the downsampling operation is given in Figure A.5. Here a highpass signal, centered at $\omega = \pi$, is subsampled by a factor $N = 2$. It is observed that the downsampled signal is the same as was the case in the first example of Figure A.4. Thus, based only on the subsampled signal, $y(m)$, we cannot determine whether the input to the subsampler was lowpass or highpass. The fact that this highpass signal, and

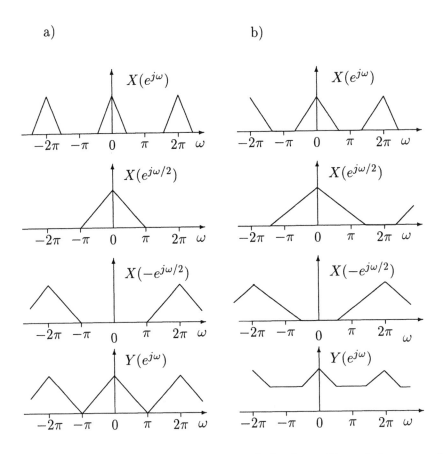

Figure A.4: *Fourier domain illustration of the formation of downsampled signals, N = 2: a) Input is appropriately bandlimited. b) Input not is appropriately bandlimited.*

in general also bandpass signals occupying the range $[p\pi/N, (p+1)\pi/N]$ with p an integer, is translated down to the baseband through subsampling is often referred to as *integer band decimation* [11].

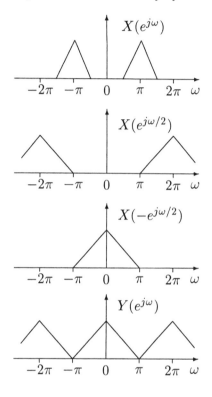

Figure A.5: *Fourier domain illustration of the formation of a subsampled highpass signal, $N = 2$.*

A.3.2 Upsampling

An N-fold upsampler is a device which inserts $N - 1$ zero valued samples between each pair of original samples in the input signal. The input/output relation is:

$$y(m) = \begin{cases} x(\frac{m}{N}), & m = 0, \pm N, \pm 2N, \dots \\ 0, & \text{otherwise.} \end{cases} \tag{A.11}$$

A block diagram along with an example input/output sequence is depicted in Figure A.6 for the case when $N = 2$. This expansion of the signal in the time domain corresponds to a compression in the frequency domain. The

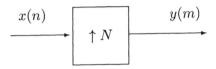

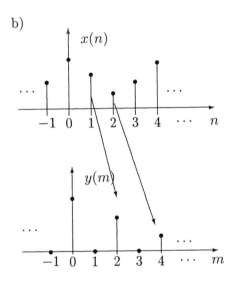

Figure A.6: *Upsampler: a) Block diagram. b) Example of operation for*
$N = 2$.

input/output relations in the z-domain and frequency domain are [11]:

$$Y(z) = X(z^N) \tag{A.12}$$

and

$$Y(e^{j\omega}) = X(e^{j\omega N}). \tag{A.13}$$

In Figure A.7 we show an example of the frequency domain behavior of
the input and the output of a 2-fold upsampler. Notice the introduction
of so-called mirror components (corresponding to replication of compressed
versions of the Fourier transform of the input signal). In most cases these

mirror components are undesirable. Therefore an upsampler is commonly followed by a lowpass filter (interpolation filter) whose task it is to remove these. The combination of an upsampler and a lowpass filter is called an *interpolator*

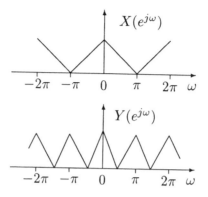

Figure A.7: *Fourier transform of signal upsampled by N = 2.*

A.3.3 The Noble Identities

In multirate systems we can gain insights as well as obtain simplifications for implementation, by performing network manipulations. The two network identities of Figure A.8 [3], which are valid when the transfer function $G(z)$ is rational, are used extensively in filter bank theory.

The equivalences in Figure A.8 above show that the order of filtering and upsampling/downsampling can be interchanged provided the z-dependency of the filter is modified. The filters containing only z as argument are easier to implement: They require less memory and trivial multiplications by zero-coefficients are automatically avoided.

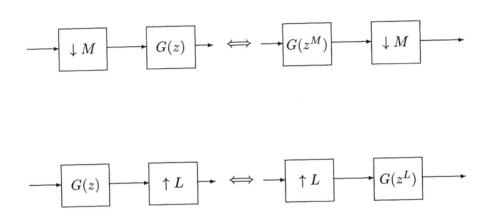

Figure A.8: *The noble identities.*

Appendix B

The Heisenberg Uncertainty Relation

Exact decomposition of a signal requires a basis for the series expansion. A sufficient condition for creating a basis for a length N discrete signal is the existence of N linearly independent functions. (See Chapter 2 for further details). This is a very mild condition which leaves much freedom for the choice of the basis functions. In Chapter 2 it is argued that series expansions can be used also for infinite length signals. One particular class of basis functions can be implemented as multirate filter banks, that is, the unit sample response of the interpolating synthesis filters and shifts of these form the basis. Also in this case much freedom is left for the design of the basis.

In practical subband coders we wish to optimize e.g. *coding gain* (see Chapter 5). The coding gain limit is reached when applying perfect filters in terms of frequency response, while the number of bands tends to infinity. The filter responses must thus be infinite, resulting in severe implications on low bit rate coding results. Visually, the typical distortion in such cases is manifested through ringing artifacts in e.g. smooth image areas close to edges. To avoid such artifacts we need filter banks with short unit sample responses. There is definitely some sort of conflict between the two above requirements.

The ringing artifact is a manifestation of the Gibbs Phenomenon. In statistical terms it becomes an annoying problem due to the nonstationarities of images. Considering optimality of the filters, their region of support should definitely not extend beyond an area for which the signal is stationary. That is, for rapidly varying statistics the unit sample responses of the filters should be short.

The simultaneous presence of short unit sample responses and good frequency selectivity of filters is impossible because there exist a Heisenberg uncertainty relation between a function and its Fourier transform. We look at this space-frequency response localization problem in the sequel.

To simplify the discussion we consider space-continuous signals representing the unit sample responses of the filters. Let $\psi_n(x)$ be the impulse response of synthesis filter no. n, and its Fourier transform be given by $\Psi_n(j\Omega)$:

$$\Psi_n(j\Omega) = \int_{-\infty}^{\infty} \psi_n(x) e^{-j\Omega x} dx, \tag{B.1}$$

where j is the imaginary unit and Ω is the frequency.

The Fourier transform pair exhibits a Heisenberg relation which makes simultaneous arbitrary fine concentration of the unit sample response and its Fourier transform impossible.

To state the Heisenberg relation for bandwidth and duration, we need to define both. Here we apply *root mean square (rms) bandwidth*, W_{rms}, and *root mean square spatial extent*, X_{rms}, which are defined as

$$W_{rms} = \left[\frac{\int_{-\infty}^{\infty} \Omega^2 |\Psi_n(j\Omega)|^2 d\Omega}{\int_{-\infty}^{\infty} |\Psi_n(j\Omega)|^2 d\Omega} \right]^{\frac{1}{2}}, \tag{B.2}$$

and

$$X_{rms} = \left[\frac{\int_{-\infty}^{\infty} x^2 |\psi_n(x)|^2}{\int_{-\infty}^{\infty} |\psi_n(x)|^2} \right]^{\frac{1}{2}}. \tag{B.3}$$

It is easily shown that

$$X_{rms} W_{rms} \geq \frac{1}{4\pi}. \tag{B.4}$$

If other definitions of bandwidth and extent are used, their product will still obey a similar inequality, although with possibly another constant.

In image coding the *visual quality* is more important than the mathematical optimality in terms of coding gain. Certain psycho-visual experiments have established that there exist mechanisms like masking in space and frequency in much the same manner as for auditory masking. In frequency domain masking, a strong signal in one frequency band will mask noise in adjacent bands. This should indicate that we should split the signal in as many bands as possible because the masking is limited to a narrow band close to the strong signal. In space domain masking, a disturbance close to a strong signal component is masked. This is typical close to edges and lines. Noise generated at an edge should remain close to the edge to exploit the masking effect. If the basis functions are long, the noise may become visible. This will be especially bad if the vicinity is a smooth area. Unfortunately, the spatial masking area is rather short. Many commonly used filter banks exceed by far this limit.

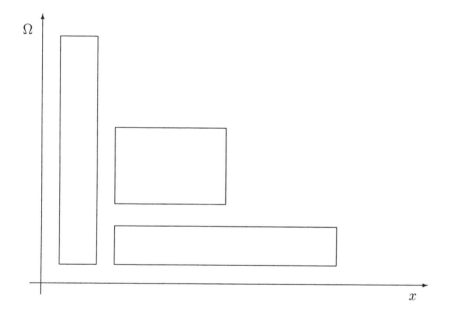

Figure B.1: *Illustration of the tradeoff between time and frequency resolution. Each rectangle has the same areas.*

In Chapter 7 we demonstrate these effects through image examples and derive practical filter banks that will try to exploit the masking effects.

To clearly state the tradeoffs between concentration in space and frequency, we can assume that the equality sign of the Heisenberg relation applies. This means that the bandwidth product is constant. In the space-frequency diagram in Figure B.1, the area in each block is equal. This is an indication of the actual partitioning of the space-frequency plane. A reversible decomposition must tile the complete space-frequency plane. From the figure we realize that such a partitioning can be done in a variety of ways.

The extent of the transients is directly given by the unit sample response length. Thus, the better frequency selectivity we wish to have, the worse the transient problem becomes.

Appendix C

Filter Coefficients

C.1 Johnston Filters

The Johnston filters [17] are commonly used as prototype lowpass filters in FIR two-band filter banks. The relation between these filters and the corresponding highpass filters is given by

$$h_{HP}(n) = (-1)^n h_{LP}(n). \tag{C.1}$$

In the frequency domain this corresponds to

$$H_{HP}(e^{j\omega}) = H_{LP}(e^{j(\omega-\pi)}), \tag{C.2}$$

which is the QMF condition. The coefficients of some of these filters are given in Table C.1.

C.2 Two-Channel Filter Banks based on Allpass Filters

In Table C.2 we list the filter coefficients of allpass filters commonly used as prototype lowpass filters in two-band allpass based filter banks [51]. The allpass filters are of first order. Branch 0 refers to the upper branch, and branch 1 refers to the lower branch in a polyphase realization of the two-channel filter bank. See Section 3.4.3 for details on the interpretation of the coefficients.

f8a	f16b	f24c	f32d
0.48998080e00	0.47734690e00	0.46864790e00	4.6367410e-01
0.69428270e-01	0.10679870e00	0.12464500e00	1.3297250e-01
-0.7065183e-01	-0.95302340e-01	-0.99878850e-01	-9.9338590e-02
0.938715e-02	-0.16118690e-01	-0.34641430e-01	-4.4524230e-02
	0.35968530e-01	0.50881620e-01	5.4812130e-02
	-0.19209360e-02	0.10046210e-01	1.9472180e-02
	-0.99722520e-02	-0.27551950e-01	-3.4964400e-02
	0.28981630e-02	-0.65046690e-03	-7.9617310e-03
		0.13540120e-01	2.2704150e-02
		-0.22731450e-02	2.0694700e-03
		-0.51829780e-02	-1.4228990e-02
		0.23292660e-02	8.4268330e-04
			8.1819410e-03
			-1.9696720e-03
			-3.9711520e-03
			2.2451390e-03

Table C.1: *Johnston type FIR QMF coefficients – the center tap is listed on top. Only one symmetric half of the filter coefficients is listed.*

Branch No.	f_2_2_06	f_2_2_08	f_2_2_smpl
0	0.1576	0.2365	0.1250
1	0.6148	0.7145	0.6125

Table C.2: *Filter coefficients for the allpass branches of some half band lowpass prototype QMFs.*

Branch No.	f_4_1_06	f_8_1_06	f_8_1_08	f_8_1_smpl
0	0.2076	0.1038	0.1663	0.125
1	0.4257	0.2078	0.3056	0.250
2	0.6772	0.3144	0.4284	0.375
3	Identity	0.4260	0.5418	0.500
4		0.5458	0.6510	0.625
5		0.6774	0.7606	0.750
6		0.8263	0.8751	0.875
7		Identity	Identity	Identity

Table C.3: *Filter coefficients for the allpass branches of fourth and eight band lowpass prototype filters.*

C.3 Parallel Filter Banks based on Allpass Filters

In Table C.3 we list filter coefficients for first-order allpass filters used in parallel IIR filter banks as discussed in Section 3.6.1. Each branch corresponds to a polyphase filter. The first filter is for a four band structure, and the others are for eight band structures.

C.4 Parallel FIR Filter Banks

In the following is a list of the coefficients of the 8-channel parallel FIR filter banks discussed in this book. The filter banks of Section 8.2 are included, as well as the DCT and the 16-tap LOT [52].

Only the first half of each filter is given, the latter half is given by even or odd symmetry. The lowpass filters have even symmetry, the first bandpass filters have odd symmetry, the next even, and so on. First we give all analysis responses: Lowpass, 1st bandpass,. . . ,highpass from left to right, then follow the synthesis responses.

32II_opt:

Analysis:

```
 0.001463 -0.002115 -0.001233  0.001414  0.000570 -0.000496  0.000559  0.001372
 0.003788 -0.000733 -0.001178  0.001311  0.000980  0.000279  0.001212  0.001665
 0.005659  0.000499 -0.001070  0.001192  0.001319  0.000958  0.001747  0.001865
 0.005409  0.000907 -0.000834  0.000937  0.001226  0.001067  0.001656  0.001617
-0.041094 -0.021863 -0.000202  0.000381 -0.006670 -0.011335 -0.010446 -0.005694
-0.052186 -0.027226  0.001992  0.005115 -0.000867 -0.005450 -0.005668 -0.002996
-0.059951 -0.033268  0.000793  0.008717  0.008468  0.006785  0.004941  0.002810
-0.071024 -0.046258 -0.009755  0.003315  0.010202  0.012839  0.010942  0.005937
-0.086731 -0.068915 -0.034049 -0.017045 -0.003548  0.004140  0.005272  0.002466
-0.101280 -0.092916 -0.063205 -0.044918 -0.027754 -0.016131 -0.009996 -0.006335
-0.109345 -0.109715 -0.086490 -0.069020 -0.050943 -0.037143 -0.026324 -0.015633
-0.093012 -0.096862 -0.080594 -0.066906 -0.052141 -0.040149 -0.029205 -0.017141
 0.320147  0.440744  0.474528  0.451816  0.419669  0.365716  0.265664  0.141891
 0.408647  0.485313  0.324811  0.063894 -0.241856 -0.444247 -0.467581 -0.293114
 0.447144  0.351034 -0.096472 -0.453138 -0.393026  0.030856  0.426813  0.411406
 0.468465  0.133234 -0.425546 -0.296930  0.334293  0.410061 -0.169436 -0.474019
```

Synthesis:

```
 0.012921  0          0          0          0          0          0          0
 0.017205  0          0          0          0          0          0          0
 0.019466  0          0          0          0          0          0          0
 0.018201  0          0          0          0          0          0          0
 0.012388  0          0          0          0          0          0          0
 0.012615  0          0          0          0          0          0          0
 0.020375  0          0          0          0          0          0          0
 0.035362  0          0          0          0          0          0          0
 0.052560  0.010906   0          0          0          0          0          0
 0.082819  0.003030   0          0          0          0          0          0
 0.126282 -0.043632 -0.021877   0          0          0          0          0
 0.185306 -0.159154  0.087693    0          0          0          0          0
 0.261039 -0.386887  0.435016 -0.427255  0.398723 -0.356641  0.259000 -0.140746
 0.318281 -0.440942  0.303884 -0.037343 -0.260467  0.453469 -0.473622  0.294039
 0.356008 -0.335291 -0.109926  0.472235 -0.399952 -0.024194  0.424353 -0.410094
 0.375174 -0.126914 -0.445155  0.305044  0.336475 -0.408152 -0.168895  0.474926
```

32II_flat:

Analysis:

```
 0.000028  0.002088  0.000139 -0.000020  0.000352  0.000571  0.000472  0.000231
 0.000089  0.003857  0.000385  0.000126  0.000812  0.001211  0.001018  0.000525
 0.000143  0.005195  0.000590  0.000263  0.001189  0.001725  0.001455  0.000761
 0.000141  0.004876  0.000577  0.000277  0.001148  0.001650  0.001394  0.000732
-0.001020 -0.024974 -0.003786 -0.002531 -0.007065 -0.009597 -0.008123 -0.004357
-0.000008 -0.029385 -0.001120  0.002187 -0.001890 -0.004820 -0.004433 -0.002309
-0.001164 -0.034322  0.000372  0.008101  0.007583  0.005721  0.004190  0.002414
-0.006001 -0.041683 -0.003964  0.008056  0.011956  0.012264  0.009950  0.005535
-0.025744 -0.070723 -0.031963 -0.014405 -0.003828  0.002052  0.003444  0.001846
```

```
-0.046469  -0.100962  -0.063877  -0.043441  -0.028637  -0.017713  -0.010740  -0.006051
-0.061879  -0.120395  -0.088862  -0.068201  -0.051352  -0.036822  -0.024821  -0.013869
-0.057842  -0.108544  -0.084370  -0.067178  -0.052493  -0.038989  -0.026849  -0.014976
 0.288821   0.445813   0.467842   0.453790   0.425803   0.362708   0.263783   0.143077
 0.368270   0.485432   0.331260   0.064013  -0.238184  -0.443800  -0.468287  -0.295602
 0.420698   0.355457  -0.086245  -0.451089  -0.393441   0.028018   0.427256   0.412125
 0.433033   0.154902  -0.436986  -0.296237   0.328029   0.414015  -0.169699  -0.471560
```

Synthesis:

```
 0.000229  0          0          0          0          0          0          0
-0.000437  0          0          0          0          0          0          0
-0.004067  0          0          0          0          0          0          0
-0.012771  0          0          0          0          0          0          0
-0.027298  0          0          0          0          0          0          0
-0.032992  0          0          0          0          0          0          0
-0.029596  0          0          0          0          0          0          0
-0.016186  0          0          0          0          0          0          0
 0.007189   0.004228  0          0          0          0          0          0
 0.042813  -0.005951  0          0          0          0          0          0
 0.094352  -0.051420  -0.007181  0          0          0          0          0
 0.165048  -0.163564   0.098609  0          0          0          0          0
 0.256445  -0.387095   0.433843  -0.433098   0.411242  -0.347341   0.254662  -0.148045
 0.324106  -0.438379   0.309039  -0.041481  -0.251904   0.460919  -0.477690   0.290765
 0.368573  -0.334733  -0.099465   0.467661  -0.399638  -0.015921   0.420935  -0.416022
```

16I:

Analysis:

```
-0.216254  -0.204064  -0.015414   0.090892   0.045543  -0.031078  -0.043883  -0.019230
-0.141840  -0.210864  -0.170144  -0.049303   0.079875   0.123403   0.080780   0.028503
-0.031802  -0.098425  -0.190195  -0.245824  -0.219588  -0.123599  -0.018515   0.021430
 0.078682   0.102644   0.067054   0.008846  -0.056673  -0.116911  -0.149160  -0.110286
 0.197183   0.319544   0.364344   0.381626   0.391419   0.388237   0.343379   0.216868
 0.312270   0.448379   0.366952   0.157551  -0.112028  -0.340972  -0.442513  -0.321212
 0.429384   0.425891  -0.005008  -0.403884  -0.430905  -0.062911   0.369191   0.399227
 0.525599   0.238693  -0.418062  -0.346508   0.302485   0.440036  -0.139322  -0.425200
```

Synthesis:

```
 0.027168   0.008626  -0.058753   0.024473   0.056091  -0.053064  -0.039169   0.074033
 0.080240   0.005740  -0.163639   0.142704   0.045688  -0.153740   0.091363  -0.009521
 0.134914  -0.081740  -0.122545   0.259109  -0.232178   0.121069  -0.032123   0.000869
 0.189573  -0.225026   0.131138  -0.032819  -0.026193   0.081170  -0.137453   0.116233
 0.244171  -0.362821   0.381070  -0.374435   0.385031  -0.386747   0.339591  -0.208989
 0.298765  -0.418156   0.341248  -0.131930  -0.131196   0.348172  -0.443354   0.318822
 0.353308  -0.342651  -0.027671   0.395978  -0.425344   0.066351   0.372125  -0.406969
 0.406174  -0.136641  -0.418480   0.310510   0.306261  -0.420836  -0.140825   0.412274
```

16-tap LOT:

Analysis:

```
-0.068620 -0.075693  0.022634 -0.020613  0.037818  0.012943 -0.046888  0.050277
-0.032015 -0.085200  0.144217  0.092490  0.068536  0.129402  0.092029 -0.049314
 0.037184 -0.003875  0.149939  0.231197 -0.225392 -0.147354 -0.023092 -0.004362
 0.128540  0.151672 -0.091301 -0.038374 -0.031080 -0.078625 -0.149543  0.132397
 0.227573  0.324127 -0.368688 -0.380098  0.384674  0.362514  0.340760 -0.266202
 0.317719  0.418332 -0.338928 -0.175841 -0.128093 -0.297582 -0.438834  0.361004
 0.384938  0.378025  0.049733  0.362079 -0.421421 -0.164244  0.369817 -0.405006
 0.418867  0.202558  0.441128  0.360000  0.316430  0.456338 -0.144840  0.335076
```

The synthesis responses are obtained by time-reversing the analysis responses.

8-tap DCT:

Analysis:

```
0.353553  0.490393  0.461940  0.415735  0.353553  0.277785  0.191342  0.097545
0.353553  0.415735  0.191342 -0.097545 -0.353553 -0.490393 -0.461940 -0.277785
0.353553  0.277785 -0.191342 -0.490393 -0.353553  0.097545  0.461940  0.415735
0.353553  0.097545 -0.461940 -0.277785  0.353553  0.415735 -0.191342 -0.490393
```

The synthesis responses are obtained by time-reversing the analysis responses.

Appendix D

Original Images

In the following we show the original images used in all experiments in this book. In all cases the green-component of the RGB representation has been used. The image dimension is 512×512 pixels.

Figure D.1: *Original image "Lenna". The white frame indicates the region used for testing of blocking effects in Section 8.2.2.*

Figure D.2: *Original image "Boat".*

Figure D.3: *Original image "Kiel".*

Bibliography

[1] J. G. Proakis and D. M. Manolakis, *Digital Signal Processing: Principles, Algorithms and Applications*. New York: Macmillan Publishing Company, 2 ed., 1992.

[2] E. Dubois, "The sampling and reconstruction of time-varying imagery with application to video systems," *Proc. IEEE*, vol. 73, pp. 502–522, Apr. 1985.

[3] P. P. Vaidyanathan, *Multirate Systems and Filter Banks*. Englewood Cliffs: Prentice Hall, 1993.

[4] A. N. Akansu and R. A. Haddad, *Multiresolution Signal Decomposition*. San Diego: Academic Press, 1992.

[5] C. Shannon, "A mathematical theory of communication," *Bell Syst. Tech. J.*, vol. 27, no. 3, pp. 379–423, 1948.

[6] T. Berger, *Rate Distortion Theory*. Englewood Cliffs: Prentice-Hall, 1971.

[7] A. Gersho, *Vector Quantization and Signal Compression*. Boston: Kluwer Academic Publishers, 1992.

[8] T. A. Ramstad and S. Lepsøy, "Block-based attractor coding: Potential and comparison to vector quantization," in *Proc. Twenty-Seventh Asilomar Conference on Signals, Systems, and Computers*, Nov. 1993.

[9] D. E. Dudgeon and R. M. Mersereau, *Multidimensional Digital Signal Processing*. Englewood Cliffs: Prentice-Hall, 1984.

[10] N. S. Jayant and P. Noll, *Digital Coding of Waveforms.* Englewood Cliffs: Prentice-Hall, 1984.

[11] R. E. Crochiere and L. R. Rabiner, *Multirate Digital Signal Processing.* Englewood Cliffs: Prentice-Hall, 1983.

[12] H. Kwakernaak and R. Sivan, *Modern Systems and Signals.* Englewood Cliffs: Prentice Hall, 1991.

[13] G. Karlsson and M. Vetterli, "Extension of finite length signals for sub-band coding," *Signal Processing*, vol. 17, pp. 161–168, June 1989.

[14] M. J. T. Smith and S. L. Eddins, "Sub-band coding of images with octave band tree structures," in *Proc. Int. Conf. Acoust. Speech, Signal Proc.*, pp. 1382–1385, 1987.

[15] S. Martucci, "Signal extension and noncausal filtering for subband coding of images," in *Proc. SPIE's Visual Communications and Image Processing*, pp. 137–148, Nov. 1991.

[16] D. Esteban and C. Galand, "Application of quadarature mirror filters to split band voice coding schemes," in *Proc. Int. Conf. Acoust. Speech, Signal Proc.*, pp. 191–195, 1977.

[17] J. D. Johnston, "A filter family designed for use in quadrature mirror filter banks," in *Proc. Int. Conf. Acoust. Speech, Signal Proc.*, (Denver, CO), pp. 291–294, IEEE, 1980.

[18] M. J. T. Smith and T. P. Barnwell, "Exact reconstruction techniques for tree-structured subband coders," *IEEE Trans. Acoust., Speech, Signal Processing*, vol. ASSP-34, pp. 434–441, June 1986.

[19] M. Vetterli, "Filter banks allowing perfect reconstruction," *Signal Processing*, pp. 219–244, 1986.

[20] K. Nayebi, T. P. Barnwell, and M. J. T. Smith, "A general time domain analysis and design framework for exactly reconstructing FIR anaysis/synthesis filter banks," in *Proc. ISCAS*, pp. 2022–2025, May 1990.

[21] K. Nayebi, T. P. Barnwell, and M. J. T. Smith, "The time domain analysis and design of exactly reconstructing FIR anaysis/synthesis filter banks," in *Proc. ICASSP*, pp. 1735–1738, Apr. 1990.

[22] S. O. Aase, *Image subband coding artifacts: Analysis and remedies*. PhD thesis, The Norwegian Institute of Technology, Mar. 1993. Available as ps-file through WWW: http://www.hsr.no/tele-signal-group/www/bibliography.html.

[23] G. Strang, *Linear Algebra and its applications*. San Diego: Harcourt Brace Jovanovich Publishers, 3rd ed., 1988.

[24] T. A. Ramstad, "Analysis/synthesis filter banks with critical sampling," in *Proc. Intl. Conf. on Digital Signal Processing*, (Florence), 1984.

[25] R. Crochiere, S. Webber, and J. Flanagan, "Digital coding of speech in sub-bands," *Bell Syst. Tech. J.*, vol. 55, no. 8, pp. 1069–1085, 1976.

[26] T. A. Ramstad and O. Foss, "Subband coder design using recursive quadrature mirror filters," in *Proc. of Eur. Signal Proc. Conf. (EUSIPCO)*, pp. 747–752, 1980.

[27] T. P. Barnwell, "An experimental study of sub-band coder design incorporating recursive quadrature filters and optimum ADPCM," in *Proc. Int. Conf. Acoust. Speech, Signal Proc.*, pp. 808–811, April 1981.

[28] J. H. Rothweiler, "Polyphase quadrature filters, a new subband coding technique," in *Proc. Int. Conf. Acoust. Speech, Signal Proc.*, (Boston), pp. 1980–1983, April 1983.

[29] P. Vaidyanathan and P. Hong, "Lattice structures for optimal design and robust implementation of two–channel perfect reconstruction QMF banks," *IEEE Trans. Acoust., Speech, Signal Processing*, vol. ASSP-36, pp. 81–94, Jan. 1988.

[30] T. A. Ramstad, "IIR filterbank for subband coding of images," in *Proc. ISCAS*, pp. 827–830, 1988.

[31] T. A. Ramstad and J. H. Husøy, "Parallel complex filterbank for sub-band coding of images," in *Proc. of The Sixth International Symposium on Network, Systems and Signal Processing*, pp. 162–165, Zagreb, Jugoslavia,1989.

[32] J. H. Husøy and T. A. Ramstad, "Application of an efficient parallel IIR filter bank to image subband coding," *Signal Processing*, vol. 20, pp. 279–293, Aug. 1990.

[33] H. S. Malvar, "Extended lapped transforms: Fast algorithms and applications," in *Proc. Int. Conf. Acoust. Speech, Signal Proc.*, (Toronto, Canada), pp. 1797–1800, May 1991.

[34] R. D. Koilpillai and P. P. Vaidyanathan, "New results on cosine-modulated FIR filter banks satisfying perfect reconstruction," in *Proc. Int. Conf. Acoust. Speech, Signal Proc.*, (Toronto), pp. 1793–1796, May 1991.

[35] T. A. Ramstad and J. P. Tanem, "Cosine-modulated analysis-synthesis filterbank with critical sampling and perfect reconstruction," in *Proc. Int. Conf. Acoust. Speech, Signal Proc.*, (Toronto, Canada), pp. 1789–1792, May 1991.

[36] S. O. Aase and T. A. Ramstad, "Parallel FIR filter banks for robust subband image coding," in *Proc. ICASSP*, (Minneapolis), Apr. 1993.

[37] I. Daubechies, "Orthonormal bases of compactly supported wavelets," *Comm. on Pure and Appl. Math.*, vol. 4, no. 7, pp. 909–996, 1988.

[38] A. K. Jain, "A sinusoidal family of unitary transforms," *IEEE Trans. Pattern Anal. Machine Intell.*, vol. 1, pp. 356–365, Oct. 1979.

[39] M. G. Bellanger and J. L. Daguet, "TDM–FDM transmultiplexer: Digital polyphase and FFT," *IEEE Trans. Commun.*, vol. COM-22, pp. 1199–1205, Sept. 1974.

[40] H. G. Martinez and T. W. Parks, "A class of infinite-duration impulse response digital filters for sampling rate reductiom," *IEEE Trans. Acoust., Speech, Signal Processing*, vol. ASSP-27, pp. 154–162, Apr. 1979.

[41] T. Ramstad, "A design method for iir sample-rate reduction," in *International Conf. on Communications*, 1981.

[42] T. A. Ramstad, "Digital two-rate IIR and hybrid FIR filters for sampling rate conversion," *IEEE Trans. Commun.*, vol. COM-30, pp. 1466–1476, July 1982.

[43] J. H. McClellan and T. W. Parks, "A unified approach to the design of optimum FIR linear-phase digital filters," *IEEE Trans. Circuits, Syst.*, vol. CT-20, pp. 697–701, Nov 1973.

[44] M. Renfors and T. Saramäki, "Recursive N-th band digital filters – part I: Design and properties," *IEEE Trans. Circuits, Syst.*, vol. CAS-34, pp. 24–39, Jan. 1987.

[45] P. Vaidyanathan, S. K. Mitra, and Y. Neuvo, "A new approach to the realization of low-sensitivity IIR digital filters," *IEEE Trans. Acoust., Speech, Signal Processing*, vol. ASSP-34, pp. 350–361, Apr. 1986.

[46] S. K. Mitra and K. Hirano, "Digital all-pass networks," *IEEE Trans. Circuits, Syst.*, vol. CAS-21, pp. 207–214, Feb. 1974.

[47] P. P. Vaidyanathan, "Theory and design of M-channel maximally decimated quadrature mirror filters with arbitrary M, having the perfect reconstruction property," *IEEE Trans. Acoust., Speech, Signal Processing*, vol. ASSP-35, pp. 476–492, Apr. 1987.

[48] M. J. T. Smith and T. P. Barnwell, "A procedure for designing exact reconstruction filter banks for tree structured subband coders," in *Proc. Int. Conf. Acoust. Speech, Signal Proc.*, (San Diego), pp. 27.1.1–27.1.4, 1984.

[49] O. Herrmann and Schüssler, "Design of nonrecursive digital filters with minimum phase," *Electronics Letters*, vol. 6, pp. 207–214, May 1970.

[50] W. Givens, "Computation of plane unitary rotations transforming a general matrix to triangular form," *SIAM J. Appl. Math.*, vol. 6, pp. 26–50, 1958.

[51] J. H. Husøy, *Subband Coding of Still Images and Video*. PhD thesis, Norwegian Institute of Technology, Jan. 1991.

[52] H. S. Malvar and D. H. Staelin, "The LOT: Transform coding of images without blocking effects," *IEEE Trans. Acoust., Speech, Signal Processing*, vol. 37, pp. 553–559, Apr. 1989.

[53] G. Bonnerot and M. G. Bellanger, "Odd-time odd-frequency discrete Fourier transform for symmetric real-valued series," *Proc. IEEE*, Mar. 1976.

[54] J. Max, "Quantizing for minimum distortion," *IEEE Trans. Inform. Theory*, pp. 7–12, Mar. 1960.

[55] S. P. Lloyd, "Least squares quantization in PCM," *IEEE Trans. Inform. Theory*, vol. IT-28, pp. 129–137, Mar. 1982.

[56] N. Farvardin and J. W. Modestino, "Optimum quantizer performance for a class of non-Gaussian memoryless sources," *IEEE Trans. Inform. Theory*, vol. IT-30, pp. 485–497, May 1984.

[57] D. A. Huffman, "A method for the construction of minimum redundancy codes," *Proc. IRE*, vol. 40, pp. 1098–1101, Sept. 1952.

[58] J. Makhoul, S. Roucos, and H. Gish, "Vector quantization in speech coding," in *Proc. IEEE*, pp. 1551–1587, Nov. 1985.

[59] J. H. Conway and N. J. A. Sloane, "A lower bound on the average error of vector quantizers," *IEEE Trans. Inform. Theory*, vol. 31, pp. 106–109, Jan. 1985.

[60] Y. Linde, A. Buzo, and R. M. Gray, "An algorithm for vector quantizer design," *IEEE Trans. Commun.*, vol. COM-28, pp. 84–95, Jan. 1980.

[61] T. D. Lookabaugh and R. M. Gray, "High-resolution quantization theory and vector quantizer advantage," *IEEE Trans. Inform. Theory*, vol. 35, pp. 1020–1033, Sept. 1989.

[62] A. F. M. Smith and U. E. Makov, *Statistical Analysis of Finite Mixture Distributios*. Wiley, 1985.

[63] D. G. Luenbereger, *Introduction to Linear and Nonlinear Programming*. Reading, Massachusetts: Addison–Wesley, 1973.

[64] T. A. Ramstad, "Sub–band coder with a simple bit–allocation algorithm, – a possible candidate for digital mobile telephony?," in *Proc. Int. Conf. Acoust. Speech, Signal Proc.*, pp. 203–207, 1982.

[65] T. A. Ramstad, "Considerations on quantization and dynamic bitallocation in subband coders," in *Proc. Int. Conf. Acoust. Speech, Signal Proc.*, pp. 841–844, 1986.

[66] John M. Lervik and Tor A. Ramstad, "Optimizing a collection of entropy coders for adaption to mixture distributions," in *Submitted to IEEE International Conference on Image Processing (ICIP-95)*.

[67] A. Gersho, T. Ramstad, and I. Versvik, "Fully vector-quantized subband coding with adaptive codebook allocation," in *Proc. Int. Conf. Acoust. Speech, Signal Proc.*, vol. 1, pp. 1–4, Mar. 1984.

[68] H. Ljøen and T. A. Ramstad, "Vector excited subband code," in *Proc. of Eur. Signal Proc. Conf. (EUSIPCO)*, (Grenoble), 1988.

[69] P. H. Westerink, D. E. Boekee, J. Biemond, and J. W. Woods, "Subband coding of images using vector quantization," *IEEE Trans. Commun.*, vol. COM-36, pp. 713–719, June 1988.

[70] H. Karhunen, "Über lineare Methoden in der Wahrscheinlichkeitsrechnung," *Ann. Acad. Sci. Fenn. Ser. A.I.37*, 1947. Helsinki, Finland.

[71] H. Hotelling, "Analysis of a complex of statistical variables into principal components," *J. Educational Psychology*, pp. 417–441, 498–520, 1933.

[72] R. A. Roberts and C. T. Mullis, *Digital Signal Processing*. Reading, Massachusetts: Addison–Wesley, 1987.

[73] J. Katto and Y. Yasuda, "Performance evaluation of subband coding and optimization of its filter coefficients," in *Proc. SPIE's Visual Communications and Image Processing*, pp. 95–106, Nov. 1991.

[74] S. O. Aase and T. A. Ramstad, "On the optimality of nonunitary filters in subband coders," *IEEE Trans. Image Processing*, 1995. To be published.

[75] H. Malvar, *Signal Processing with Lapped Transforms*. Artech House, 1992.

[76] P. M. Cassereau, D. H. Staelin, and G. D. Jager, "Encoding of images on a lapped orthogonal transform," *IEEE Trans. Commun.*, vol. 37, pp. 189–193, Feb. 1989.

[77] P. H. Westerink, *Subband Coding of Images*. PhD thesis, Technical University of Delft, Oct. 1989.

[78] A. V. Oppenheim and R. W. Schafer, *Digital Signal Processing*. Englewood Cliffs: Prentice-Hall, 1975.

[79] S. J. Orfanidis, *Optimum Signal Processing: An Introduction*. New York: Macmillan Publishing Company, 1985.

[80] J. W. Woods and S. D. O'Neill, "Subband coding of images," *IEEE Trans. Acoust., Speech, Signal Processing*, vol. ASSP-34, pp. 1278–1288, Oct. 1986.

[81] B. Girod, F. Hartung, and U. Horn, "Subband image coding," in *Subband Coding* (M. Smith and A. Akansu, eds.), Kluwer Adademic Publishers, 1995.

[82] R. Ansari, A. Fernandez, and S. H. Lee, "HDTV subband/DCT coding using IIR filter banks: Coding strategies," in *Proc. SPIE's Visual Communications and Image Processing*, pp. 1291–1302, 1989.

[83] A. N. Netravali and B. G. Haskell, *Digital Pictures*. New York: Plenum Press, 1988.

[84] H. Gharavi and A. Tabatabai, "Sub-band coding of monochrome and color images," *IEEE Trans. Circuits, Syst.*, vol. CAS-35, pp. 207–214, Feb. 1988.

[85] W.-H. Chen and W. Pratt, "Scene adaptive coder," *IEEE Trans. Commun.*, vol. COM-32, pp. 225–232, Mar. 1984.

[86] CCITT Study Group XV, Working Party XV, "Recommendation H.261 - video codec for audiovisual services at p x 64 kbit/s," *Study Group XV – Report R 37*, July 1990.

[87] A. K. Jain, *Fundamentals of digital image processing*. Englewood Cliffs: Prentice-Hall, 1989.

[88] J. S. Lim, *Two-dimensional signal and image processing*. Englewood Cliffs: Prentice-Hall, 1990.

[89] C.-S. Kim, M. J. T. Smith, and R. M. Mersereau, "An improved SBC/VQ scheme for color image coding," in *Proc. Int. Conf. Acoust. Speech, Signal Proc.*, pp. 1941–1944, 1989.

[90] C.-S. Kim, J. Bruder, M. J. T. Smith, and R. M. Mersereau, "Subband coding of color images using finite state vector quantization," in *Proc. Int. Conf. Acoust. Speech, Signal Proc.*, pp. 753–756, 1988.

[91] S. O. Aase and T. A. Ramstad, "Ringing reduction in low bit rate image subband coding using projection onto a space of paraboloids," *Signal Processing: Image Communication*, 1993.

[92] H. Caglar, Y. Liu, and A. Akansu, "Statistically optimized PR-QMF design," in *Proc. SPIE's Visual Communications and Image Processing*, pp. 86–94, Nov. 1991.

[93] T. Kronander, *Some aspects of perception based image coding*. PhD thesis, Linköping University, Jan. 1989.

[94] R. H. Bamberger and M. J. T. Smith, "A multi-rate filter bank pre-processor for image compression," in *IEEE Southeastcon*, pp. 890–895, 1989.

[95] S. O. Aase and T. A. Ramstad, "Some fundamental experiments in subband coding of images," in *Proc. SPIE's Visual Communications and Image Processing*, (Boston), pp. 734–744, Nov. 1991.

[96] A. Cantoni and P. Butler, "Eigenvalues and eigenvectors of symmetric centrosymmetric matrices," *Linear Algebra Applications*, no. 13, pp. 275–288, 1976.

[97] W. H. Press, B. P. Flannery, S. A. Teukolsky, and W. T. Vetterling, *Numerical Recipes in C*. Cambridge: Cambridge University Press, 1988.

[98] H. S. Malvar, "Lapped transforms for efficient transform/subband coding," *IEEE Trans. Acoust., Speech, Signal Processing*, vol. 38, pp. 969–978, June 1990.

[99] K. Nayebi, T. Barnwell, and M. Smith, "Analysis-synthesis systems with time-varying filter bank structures," in *Proc. Int. Conf. Acoust. Speech, Signal Proc.*, pp. IV–617 – IV–620, 1992.

[100] C. Herley, J. Kovacevic, K. Ramchandran, and M. Vetterli, "Tilings of the time-frequency plane: Construction of arbitrary orthogonal bases and fast tiling transforms," *IEEE Trans. Signal Processing*, vol. 41, pp. 3341–3359, Dec 1993.

[101] J. H. Husøy and S. O. Aase, "Image subband coding with adaptive filter banks," in *Proc. SPIE's Visual Communications and Image Processing*, (Boston), pp. 2 – 11, Nov. 1992.

[102] M. Bever, J. H. Husøy, S. O. Aase, and T. A. Ramstad, "Reduction of coding artifacts in subband coders using adaptive filter banks," in *Proc. SPIE International symposium on optical instrumentation and applied science: Applications of Digital Image Processing XVI*, (San Diego), pp. 13 – 25, July 1993.

[103] R. Sørhus and J. H. Husøy, "Image subband coding with adaptive filter banks: Automatic filter selection," in *Proc. of Eur. Signal Proc. Conf. (EUSIPCO)*, (Edinburgh, UK), pp. 1230–1233, Sept. 1994.

[104] R. C. Gonzales and P. Wintz, *Digital Image Processing*. Reading, Massachusetts: Addison Wesley, 2 ed., 1987.

[105] Tor A. Ramstad and John M. Lervik, "Efficient use of bit resources in frequency domain codecs," in *Proc. Nordic Signal Processing Symposium (NORSIG -94)*, (Ålesund, Norway), pp. 2–7, NORSIG, June 1994. Invited paper.

[106] John M. Lervik and Tor A. Ramstad, "Optimal entropy coding in image subband coders," in *Proc. First IEEE International Conference on Image Processing (ICIP-94)*, (Austin, Texas), IEEE, Nov. 1994.

[107] T. Ishiguro and K. Iinuma, "Television bandwidth compression transmission by motion-compensated interframe coding," *IEEE Communications Magazine*, vol. 20, pp. 24–30, Nov. 1982.

[108] F. W. Mounts, "A video encoding system with conditional picture-element replenishment," *Bell Syst. Tech. J.*, vol. 48, pp. 2545–2554, Sept. 1969.

[109] T. R. Hsing, "Packet video: Video communication in broadband networks." Course notes of short course held during the Visual Communications and Image Processing conference in Lausanne, 1990.

[110] B. G. Haskell, "Frame replenishment coding of television," in *Image Transmission Techniques* (W. K. Pratt, ed.), Academic Press, 1979.

[111] M. Rabbani and R. W. Jones, *Digital Image Compression Techniques*. Bellingham, Washington: SPIE Press, 1991.

[112] B. G. Haskell, "Entropy measurements for nonadaptive and adaptive, frame-to-frame, linear predictive coding of video-telephone signals," *Bell Syst. Tech. J.*, vol. 54, pp. 1155–1174, July – Aug. 1975.

[113] P. Pirsch, "Adaptive intra-interframe DPCM coder," *Bell Syst. Tech. J.*, vol. 61, pp. 747–764, May – June 1982.

[114] A. Habibi, "An adaptive strategy for hybrid image coding," *IEEE Trans. Commun.*, vol. COM-29, pp. 1736–1740, Dec. 1981.

[115] H. Gharavi, "Subband coding algorithms for video applications: Videophone to HDTV-conferencing," *IEEE Trans. Circuits, Syst. for Video Tech.*, vol. 1, pp. 174–183, June 1991.

[116] ISO-IEC/JTCI/SC2/WG8, "MPEG video simulation model one (sm1), document MPEG 90/041," March 26 1990.

[117] A. N. Netravali and J. D. Robbins, "Motion–compensated television coding: Part I," *Bell Syst. Tech. J.*, vol. 58, pp. 631–670, Mar. 1979.

[118] J. R. Jain and A. K. Jain, "Displacement measurement and its application in interframe image coding," *IEEE Trans. Commun.*, vol. COM-29, pp. 1799–1808, Dec. 1981.

[119] J. W. Woods and T. Naveen, "Subband encoding of video sequences," in *Proc. SPIE's Visual Communications and Image Processing*, pp. 724–732, 1989.

[120] T. Koga, K. Iinuma, Y. Iijima, A. Hirano, and T. Ishiguro, "Motion-compensated interframe coding for video conferencing," in *Proc. NTC*, pp. G5.3.1–G5.3.5, 1981.

[121] Y. Ninomiya and Y. Ohtsuka, "A motion-compensated interframe coding scheme for television pictures," *IEEE Trans. Commun.*, vol. COM-30, pp. 201–211, Jan. 1982.

[122] R. Srinivasan and K. R. Rao, "Predictive coding based on efficient motion estimation," *IEEE Trans. Commun.*, vol. COM-33, pp. 888–896, Aug. 1985.

[123] S. C. Kwatra, C.-M. Lin, and W. A. Whyte, "An adaptive algorithm for motion compensated color image coding," *IEEE Trans. Commun.*, vol. COM-35, pp. 747–754, July 1987.

[124] K. Yang, L.Wu, H. Chong, and M. Sun, "VLSI implementation of motion compensation full search block-matching algorithm," in *Proc. SPIE's Visual Communications and Image Processing*, pp. 892–899, 1988.

[125] M. Nickel and J. H. Husøy, "A hybrid coder for image sequences using detailed motion estimates," in *Proc. SPIE's Visual Communications and Image Processing*, (Boston), pp. 963–971, Nov. 1991.

[126] P. H. Westerink, J. Biemond, and F. Muller, "Subband coding of image sequences at low bit rates," *Signal Processing: Image Communication*, vol. 2, pp. 441–448, Dec. 1990.

[127] J. Biemond, B. P. Thieme, and D. Boekee, "Subband coding of moving images using hierarchical motion estimation and vector quantization," in *Abstracts of Int. workshop on 64 KBPS coding of moving video*, pp. 3.2–3.3, 1988.

[128] M. H. Fadzil and T. J. Dennis, "Sample selection in subband vector quantization," in *Proc. Int. Conf. Acoust. Speech, Signal Proc.*, pp. 2085–2088, 1990.

[129] M. H. A. Fadzil and T. J. Dennis, "Video subband VQ coding at 64 kbit/s using short-kernel filter banks with an improved motion estimation technique," *Signal Processing: Image Communication*, vol. 3, pp. 3–22, Feb. 1991.

[130] A. N. Akansu and M. S. Kadur, "Subband coding of video with adaptive vector quantization," in *Proc. Int. Conf. Acoust. Speech, Signal Proc.*, pp. 2109–2112, 1990.

[131] J. H. Husøy, "Low complexity subband coding of still images and video," *Optical Engineering*, vol. 30, pp. 904–911, July 1991.

[132] J. H. Husøy, H. Grønning, and T. A. Ramstad, "Subband coding of video employing efficient recursive filter banks and advanced motion compensation," in *Proc. SPIE's Visual Communications and Image Processing*, pp. 546–557, Oct. 1990.

[133] H. Gharavi, "Differential subband coding of video signals," in *Proc. Int. Conf. Acoust. Speech, Signal Proc.*, pp. 1819–1822, May 1989.

[134] H. Gharavi, "Subband coding of video signals," in *Subband Image Coding* (J. W. Woods, ed.), ch. 6, pp. 229–272, Boston: Kluwer Academic Publishers, 1991.

[135] R. Forcheimer and T. Kronander, "Image coding – from waveforms to animation," *IEEE Trans. Acoust., Speech, Signal Processing*, vol. 37, pp. 2008–2023, Dec. 1989.

[136] C. I. Podilchuk, N. S. Jayant, and P. Noll, "Sparse codebooks for the quantization of non-dominant sub-bands in image coding," in *Proc. Int. Conf. Acoust. Speech, Signal Proc.*, pp. 2101–2104, 1990.

[137] C. I. Podilchuk and N. Farvardin, "Perceptually based low bit rate video coding," in *Proc. Int. Conf. Acoust. Speech, Signal Proc.*, pp. 2837–2840, 1991.

[138] G. Karlsson and M. Vetterli, "Packet video and its integration into the network architechture," *IEEE J. Select. Areas Commun.*, vol. 7, pp. 739–751, June 1989.

[139] G. Karlsson, *Subband coding for packet video*. PhD thesis, Colombia University, 1989.

[140] M. Gilge, "A high quality videophone coder using hierarchical motion estimation and structure coding of the prediction error," in *Proc. SPIE's Visual Communications and Image Processing*, pp. 864–874, 1988.

[141] J. H. Husøy and F. Møretrø, "A software implementation of an efficient image subband coder," in *Proc. SPIE Symposium on Electronic Imaging*, (San Jose), pp. 963–971, Feb. 1992.

[142] D. J. L. Gall and A. Tabatabai, "Subband coding of digital images using symmetric short kernel filters and arithmetic coding techniques," in *Proc. Int. Conf. Acoust. Speech, Signal Proc.*, pp. 761–764, Apr. 1988.

[143] Joint Photographic Experts Group ISO/IEC, JTC/SC/WG8, CCITT SGVIII, "JPEG technical specifications, revision 5," *Report JPEG-8-R5*, Jan. 1990.

[144] CCITT Study Group XV, Working Party XV/4, "Description of ref. model 7 (rm7), document no. 446," *Study Group XV*, Jan. 21 1989.

[145] H. Schiller and B. Chaudhuri, "Efficient coding of side information in a low bitrate hybrid image coder," *Signal Processing*, vol. 19, pp. 61–73, Jan. 1990.

[146] W. Y. Choi and R.-H. Park, "Motion vector coding with conditional transmission," *Signal Processing*, vol. 18, pp. 259–267, Nov. 1989.

[147] A. N. Akansu, J. H. Chen, and M. Kadur, "Lossless compression of block motion information in motion compensated video coding," in *Proc. SPIE's Visual Communications and Image Processing*, pp. 30–38, Nov. 1989.

[148] R. Gallager, "Variations on a theme by Huffman," *IEEE Trans. Inform. Theory*, vol. IT-24, pp. 668–674, Nov. 1978.

[149] A. von Brandt, "Subband coding of videoconference signals using quadrature mirror filters," in *Applied signal processing, Proc. of IASTED Int. Symp.*, pp. 212–215, 1985.

[150] A. von Brandt, "Motion esimation and subband coding using quadrature mirror filters," in *Proc. of Eur. Signal Proc. Conf. (EUSIPCO)*, pp. 829–832, 1986.

[151] H. Jozawa and H. Watanabe, "Video coding with motion compensated subband/wavelet decomposition," in *Proc. SPIE's Visual Communications and Image Processing*, (Boston), pp. 765 – 774, Nov. 1992.

[152] R. W. Young and N. G. Kingsbury, "Video compression using lapped transforms for motion estimation/compensation and coding," in *Proc. SPIE's Visual Communications and Image Processing*, (Boston), pp. 276 – 288, Nov. 1992.

[153] J. Roese, W. Pratt, and G. Robinson, "Interframe cosine transform image coding," *IEEE Trans. Commun.*, pp. 1329–1339, Nov. 1977.

[154] R. Ansari and D. LeGall, "Advanced television coding using exact reconstruction filter banks," in *Subband Image Coding*, ch. 7, pp. 273–318, Boston: Kluwer Academic Publishers, 1991.

[155] L. Vandendorpe, "Hierarchical transform and subband coding," *Signal Processing: Image Communication*, vol. 4, no. 3, pp. 253–255, 1992.

[156] M. Pecot, P. J. Tourtier, and Y. Thomas, "Compatible coding of television images: Part 2. compatible system," *Signal Processing: Image Communication*, vol. 2, pp. 259–268, Oct. 1990.

[157] R. Ansari, S. H. Lee, and L. Wu, "Subband image coding using IIR filters," in *Proc. 1988 Conf. Inform. Sci. and Syst.*, pp. 16–21, 1988.

[158] J. W. Woods and T. Naveen, "A filter based bit allocation scheme for subband compression of HDTV," *IEEE Trans. Image Processing*, vol. 1, pp. 436 – 439, July 1992.

[159] J. Biemond, F. Bosveld, and R. L. Lagendijk, "Hierarchical subband coding of HDTV in BISDN," in *Proc. Int. Conf. Acoust. Speech, Signal Proc.*, pp. 2113–2116, 1990.

[160] F. Bosveld, R. Lagendijk, and J. Biemond, "Performance evaluation of hierarchical coding schemes for HDTV," in *Proc. of Eur. Signal Proc. Conf. (EUSIPCO)*, pp. 801–804, Sept. 1990.

[161] F. Bosveld, R. L. Lagendijk, and J. Biemond, "Hierarchical coding of HDTV," *Signal Processing: Image Communication*, vol. 4, pp. 195 – 225, Sept. 1992.

[162] F. Bosveld, R. L. Lagendijk, and J. Biemond, "Compatible spatio-temporal subband encoding of HDTV," *Signal Processing*, vol. 28, pp. 271 – 290, Sept. 1992.

[163] P. J. Tourtier, M. Pécot, and J. F. Vial, "Motion compensated subband coding schemes for compatible high definition TV," *Signal Processing: Image Communication*, vol. 4, pp. 325 – 344, Sept. 1992.

[164] J. Mau, E. Boruguignat, and H. Amor, "Sub-band source coding for HDTV," *EBU TECHNICAL REVIEW*, pp. 34–44, Spring 1992.

[165] H. M. Hang, R. Leonardi, B. G. Haskell, R. L. Schmidt, H. Bheda, and J. Othmer, "Digital HDTV compression using parallel motion-compensated transform coders," *IEEE Trans. Circuits, Syst. for Video Tech.*, vol. 1, pp. 210–221, June 1991.

[166] T. Hidaka and K. Ozawa, "ISO/IEC JTC1 SC29/WG11; Report on MPEG-2 subjective assessment at Kurihama," *Signal Processing: Image Communication*, no. 1-2, pp. 127–157, 1993.

[167] *Signal Processing: Image Communication*, no. 1-2, 1993.

[168] G. Schamel and J. D. Lameillieure, "Subband based TV coding," *Signal Processing: Image Communication*, no. 1-2, pp. 105–118, 1993.

[169] S.-M. Lei, T.-C. Chen, and K.-H. Tzou, "Subband HDTV coding using high-order conditional statistics," *IEEE J. Select. Areas Commun.*, vol. SAC-11, pp. 65–76, Jan. 1993.

Index

Printed and bound by CPI Group (UK) Ltd, Croydon, CR0 4YY

22/10/2024

01777749-0002